OPPORTUNITIES FOR THE EMPLOYMENT OF SIMULATION IN U.S. AIR FORCE TRAINING ENVIRONMENTS

A WORKSHOP REPORT

Committee on Opportunities for the Employment of Simulation
in U.S. Air Force Training Environments: A Workshop

Air Force Studies Board

Division on Engineering and Physical Sciences

NATIONAL RESEARCH COUNCIL
OF THE NATIONAL ACADEMIES

THE NATIONAL ACADEMIES PRESS
Washington, D.C.
www.nap.edu

THE NATIONAL ACADEMIES PRESS **500 Fifth Street, NW** **Washington, DC 20001**

NOTICE: The project that is the subject of this report was approved by the Governing Board of the National Research Council, whose members are drawn from the councils of the National Academy of Sciences, the National Academy of Engineering, and the Institute of Medicine. The members of the committee responsible for the report were chosen for their special competences and with regard for appropriate balance.

This is a report of work supported by Grant FA9550-12-1-0413 between the U.S. Air Force and the National Academy of Sciences. Any opinions, findings, or conclusions expressed in this publication are those of the author(s) and do not necessarily reflect the view of the organizations or agencies that provided support for the project.

International Standard Book Number-13: 978-0-309-36813-1
International Standard Book Number-10: 0-309-36813-8

Limited copies of this report are available from:

Air Force Studies Board
National Research Council
500 Fifth Street, NW
Washington, DC 20001
(202) 334-3111

Additional copies are available from:

The National Academies Press
500 Fifth Street, NW
Keck 360
Washington, DC 20001
(800) 624-6242 or (202) 334-3313
http://www.nap.edu

Printed in the United States of America

THE NATIONAL ACADEMIES

Advisers to the Nation on Science, Engineering, and Medicine

The **National Academy of Sciences** is a private, nonprofit, self-perpetuating society of distinguished scholars engaged in scientific and engineering research, dedicated to the furtherance of science and technology and to their use for the general welfare. Upon the authority of the charter granted to it by the Congress in 1863, the Academy has a mandate that requires it to advise the federal government on scientific and technical matters. Dr. Ralph J. Cicerone is president of the National Academy of Sciences.

The **National Academy of Engineering** was established in 1964, under the charter of the National Academy of Sciences, as a parallel organization of outstanding engineers. It is autonomous in its administration and in the selection of its members, sharing with the National Academy of Sciences the responsibility for advising the federal government. The National Academy of Engineering also sponsors engineering programs aimed at meeting national needs, encourages education and research, and recognizes the superior achievements of engineers. Dr. C. D. Mote, Jr., is president of the National Academy of Engineering.

The **Institute of Medicine** was established in 1970 by the National Academy of Sciences to secure the services of eminent members of appropriate professions in the examination of policy matters pertaining to the health of the public. The Institute acts under the responsibility given to the National Academy of Sciences by its congressional charter to be an adviser to the federal government and, upon its own initiative, to identify issues of medical care, research, and education. Dr. Victor J. Dzau is president of the Institute of Medicine.

The **National Research Council** was organized by the National Academy of Sciences in 1916 to associate the broad community of science and technology with the Academy's purposes of furthering knowledge and advising the federal government. Functioning in accordance with general policies determined by the Academy, the Council has become the principal operating agency of both the National Academy of Sciences and the National Academy of Engineering in providing services to the government, the public, and the scientific and engineering communities. The Council is administered jointly by both Academies and the Institute of Medicine. Dr. Ralph J. Cicerone and Dr. C. D. Mote, Jr., are chair and vice chair, respectively, of the National Research Council.

www.national-academies.org

COMMITTEE ON OPPORTUNITIES FOR THE EMPLOYMENT OF SIMULATION IN U.S. AIR FORCE TRAINING ENVIRONMENTS: A WORKSHOP

DONALD C. FRASER (NAE), Charles Stark Draper Laboratory (retired), *Co-Chair*
RAYMOND E. JOHNS, JR., FlightSafety International, *Co-Chair*
ROBERT R. ALLARDICE, Allardice Enterprises
JOHN-PAUL B. CLARKE, Georgia Institute of Technology
JOHN D.W. CORLEY, U.S. Air Force (retired)
STEPHEN D. DETRO, Lockheed Martin Mission Systems and Training
PAMELA A. DREW, Exelis, Inc.
RICHARD V. REYNOLDS, The VanFleet Group, LLC
HARRY M. ROBINSON, Veterans Health Administration
MICHAEL J. ZYDA, University of Southern California

Staff

JOAN FULLER, Director, Air Force Studies Board
CARTER W. FORD, Program Officer
MARGUERITE E. SCHNEIDER, Administrative Coordinator
DIONNA C. ALI, Research Assistant
NORMAN M. HALLER, Consultant

AIR FORCE STUDIES BOARD

GREGORY S. MARTIN, GS Martin Consulting, *Chair*
DONALD C. FRASER (NAE), Charles Stark Draper Laboratory (retired), *Vice Chair*
BRIAN A. ARNOLD, Peachtree City, Georgia
ALLISON ASTORINO-COURTOIS, National Security Innovations, Inc.
CLAUDE M. BOLTON, JR., Defense Acquisition University
STEVEN R.J. BRUECK, University of New Mexico
THOMAS J. BURNS, Great Falls, Virginia
FRANK J. CAPPUCCIO, Cappuccio and Associates, LLC
BLAISE J. DURANTE, U.S. Air Force (retired)
BRENDAN B. GODFREY, University of Maryland at College Park
MICHAEL A. HAMEL, Lockheed Martin Space Systems Company
DANIEL E. HASTINGS, Massachusetts Institute of Technology
RAYMOND E. JOHNS, JR., FlightSafety International
PAUL G. KAMINSKI (NAE), Technovation, Inc.
ROBERT H. LATIFF, R. Latiff Associates
NANCY G. LEVESON (NAE), Massachusetts Institute of Technology
MARK J. LEWIS, Institute for Defense Analyses Science and Technology Policy Institute
ALEX MILLER, University of Tennessee
RICHARD V. REYNOLDS, The VanFleet Group, LLC
STARNES E. WALKER, University of Delaware
DEBORAH WESTPHAL, Toffler Associates
DAVID A. WHELAN (NAE), Boeing Defense, Space, and Security
REBECCA WINSTON, Winston Strategic Management Consulting

Staff

JOAN FULLER, Director
ALAN H. SHAW, Deputy Director
DIONNA C. ALI, Research Assistant
GREGORY EYRING, Senior Program Officer
CARTER W. FORD, Program Officer
CHRIS JONES, Financial Manager
MARGUERITE E. SCHNEIDER, Administrative Coordinator

Acknowledgment of Reviewers

This report has been reviewed in draft form by individuals chosen for their diverse perspectives and technical expertise, in accordance with procedures approved by the National Research Council's (NRC's) Report Review Committee. The purpose of this independent review is to provide candid and critical comments that will assist the institution in making its published report as sound as possible and to ensure that the report meets institutional standards for objectivity, evidence, and responsiveness to the study charge. The review comments and draft manuscript remain confidential to protect the integrity of the deliberative process. We wish to thank the following individuals for their review of this report:

Bimal Aponso, National Aeronautics and Space Administration,
R. Stephen Berry (NAS), Professor Emeritus, The University of Chicago,
Thomas E. Romesser (NAE), Independent Consultant, and
Jeffery A. Schroeder, Federal Aviation Administration.

Although the reviewers listed above have provided many constructive comments and suggestions, they were not asked to endorse the views presented at the workshop, nor did they see the final draft of the workshop report before its release. The review of this workshop report was overseen by Robert J. Hermann (NAE), Independent Consultant. Appointed by the NRC, he was responsible for making certain that an independent examination of this workshop report was carried out in accordance with institutional procedures and that all review comments were carefully considered. Responsibility for the final content of this report rests entirely with the committee and the institution.

Contents

Acronyms

ACC	Air Combat Command
AETC	Air Education and Training Command
AFAMS	Air Force Agency for Modeling and Simulation
AFGSC	Air Force Global Strike Command
AFI	Air Force Instruction
AFPD	Air Force Policy Directive
AFRL	Air Force Research Laboratory
AFSB	Air Force Studies Board
AFSOC	Air Force Special Operations Command
AMC	Air Mobility Command
ATD	advanced technology demonstration
C2	command and control
CAF	Combat Air Forces
CCDR	Contractor Critical Design Review
CFL	Core Function Lead
CLS	Contractor Lifecyle Support
COCOM	combatant command
CSAF	Chief of Staff of the Air Force
DMO	distributed mission operations
DoD	Department of Defense
FAA	Federal Aviation Administration
HAF	Headquarters Air Force
HPW	Human Performance Wing
I-LVC	integrated-live, virtual, constructive (LVC)
LPTA	lowest price, technically acceptable
LVC	live, virtual, constructive (training)
MAF	Mobility Air Forces
MAJCOM	major command
MWS	major weapon system

NAE	Naval Aviation Enterprise
NASA	National Aeronautics and Space Administration
NRC	National Research Council
NSA	National Security Agency
OPLAN	operational plan
RGM	rapid global mobility
SECAF	Secretary of the Air Force
TOR	terms of reference
TSRA	Training System Requirement Analysis
TSSC	Technical Support Services Contract
UT-IISC	University of Toledo Interprofessional Immersive Simulation Center
VIR	virtual immersive reality

Overview

CONTEXT FOR THE WORKSHOP

Simulators currently provide an alternative to aircraft when it comes to training requirements, both for the military and for commercial airlines. For the U.S. Air Force, in particular, simulation for training offers a cost-effective way, and in many instances a safer way in comparison with live flying, to replicate real-world missions. Current technical issues related to simulation for training include simulation fidelity and multi-level security, among others, which will need to be addressed in order for the Air Force to take full advantage of this technology.

In this context, the Deputy Assistant Secretary of the Air Force for Science, Technology, and Engineering requested that the Air Force Studies Board of the National Academies' National Research Council (NRC) undertake a 3-day workshop to (1) examine how simulation is currently used in military services, private industry, and other government agencies, such as the Federal Aviation Administration and NASA; (2) compare alternative uses to current Air Force practices to identify areas where the Air Force can benefit by adopting such practices; (3) examine how current and future technologies will allow the Air Force to gain even more benefit from simulation; and (4) examine how the combination of live training, virtual training in simulators, and constructive/computer generated entities can improve aircrew training. Regarding topics 2 through 4, the areas where the Air Force can benefit will be grouped into two categories: (1) areas that enhance and/or augment the learning process and (2) areas that may be used as a substitute for some training requirements with operational systems.

A committee of experts was appointed by the NRC in October 2014. The workshop was held on November 17-19, 2014, in Dayton, Ohio. Speakers were asked to respond to the following questions:

1. What are you doing now with simulation?
2. What are your current limitations?
3. What would you like to be able to do?
4. What technologies, approaches, and techniques do you think have promise to help make your desires in #3 possible?

The scope of the workshop focused on technologies and practices that could be applicable to high-end aircraft simulations. Thus, representatives of the U.S. Navy were invited to present on the uses of simulation for training by the Naval Aviation Enterprise, while the representatives of the U.S. Army, which is a fairly sophisticated user of simulation, were not present.

RECURRING THEMES ARISING DURING THE WORKSHOP

During the course of the 3-day workshop, common messages, or themes, appeared as a result of various presentations and resulting dialog among the participants. These themes are listed below along with the names of the participants who identified the common message. Details underlying each theme are found in the body of the report. *The report summarizes the views expressed by individual workshop participants. While the committee is responsible for the overall quality and accuracy of the report as a record of what transpired at the workshop, the views contained in this section and in the rest of the report are not necessarily those of all workshop participants, the committee, or the National Research Council.*

1. For current and future warfighters to be operationally ready on a continuous basis, realistic training in a simulated environment is critical. For Air Combat Command, in particular, training in the live (L) construct linked to Virtual Constructive (VC) is imperative for mission success. For Air Mobility Command training, VC is critical, but its requirements are somewhat fewer with regard to linking to the L environment. With respect to live, virtual, and constructive training (LVC), Air Force Special Operations Command's requirements are between Air Combat Command and Air Mobility Command (Ray Johns, John Corley) (see Chapter 1).[1]
2. Establishing stated requirements for live, virtual, and constructive training as well as implementing a LVC training strategy, capability, and governance model could greatly benefit the Air Force across its full range of missions. This undertaking will likely mean establishing a durable understanding of LVC training's relative worth compared to other components of readiness (Ray Johns, Donald Fraser) (see Chapter 1).
3. Currently, LVC training efforts are evolving in a largely ad hoc, stovepiped, and somewhat inefficient fashion. This situation suggests Air Force consideration of a different architectural approach that would be world-centric—open, pluggable, and playable—rather than platform- and contractor-proprietary-centric. This world-centric construct would contain common elements and live data, such as weather, terrain, threats, with an array of specific simulation platforms around the periphery drawing information from the common databases as opposed to utilizing their own proprietary database (Pamela Drew, Harry Robinson) (see Chapter 3).
4. There are indications that some elements of the Air Force simulation architecture currently have these world-centric enterprise characteristics, so continued pursuit of an enterprise-level solution to LVC training could be very beneficial (Pamela Drew, Harry Robinson) (see Chapter 3).
5. Advances in technology and increasingly complex user needs have led to LVC training as the primary way to train for some missions (Robert Allardice) (see Chapter 3).
6. Substantial benefits could accrue to the Air Force if it relied on open systems and acquired data rights as the model when procuring new systems. Enforcing compliance to more interoperable, related standards could lead to a "plug and play" environment (Pamela Drew, Michael Zyda) (see Chapter 3).
7. Research into the "science of learning" is indicating that young people, who have considerable computer skills compared to previous generations, learn in very different ways compared to older generations. Future architectures and systems would benefit by taking this knowledge into account (adaptive learning) (Donald Fraser, Steve Detro) (see Chapter 3).

[1] *Simulation* is a method for implementing a model over time. *Live simulations* are simulations involving real people operating real systems. *Virtual simulations* are simulation involving real people operating simulated systems or in simulated environments. *Constructive simulations* are simulations that involve simulated people operating simulated systems. (Real people may simulate the simulation by inputs, but they are not involved in determining the outcome) (see Old Dominion University, Modeling & Simulation Course MSIM 695-JAN 2003, Introduction to Combat Modeling and Simulation, Norfolk, Va.).

1

Air Force Simulation Needs

INTRODUCTION

Simulation for training has long been a central part of U.S. aviation. Pilots were first trained on the famous Link Trainer starting in 1934, when the Army Air Corps bought six Link Trainers to assist in training pilots to fly at night and in bad weather relying only on instruments. The World War II era brought orders for thousands of Link Trainers from the United States and many foreign countries. Although Army Air Forces aviation cadets flew various trainer aircraft, virtually all took blind-flying instruction in a Link Trainer.[1]

Today, commercial airline pilots are trained and certified by the Federal Aviation Administration (FAA) for flight operations almost exclusively on simulators. Advances in computer technologies, particularly virtual reality used for gaming, have provided new opportunities for using simulation to approach reality. Simulation techniques known as live, virtual, and constructive (LVC) have been under study by Air Force researchers since the early 1990s. During a visit by the National Academies' Air Force Studies Board (AFSB) in 2011 to Scott Air Force Base, General Ray Johns, then commander of the U.S. Air Force's Air Mobility Command (AMC), suggested, as one of several study topics, a look at migrating additional aircrew training to simulators in a resource-constrained environment. Later actions by the Deputy Assistant Secretary of the Air Force for Science, Technology, and Engineering and the AFSB led to National Research Council approval of terms of reference (TOR) for this workshop and subsequent appointment of the members of the Committee on Opportunities for the Employment of Simulation in U.S. Air Force Training Environments: A Workshop (see Box 1-1).[2]

The workshop opened with introductions of the large number of participants and guests, several dozen in all. The committee co-chairs thanked the many attendees and noted that this workshop represented both a challenge and an opportunity to assist the Air Force in moving forward with simulation capabilities that could benefit the service in all aspects of its mission. They also established that the greatest benefit of a workshop like this would be the dialog, discourse, and discussions resulting from the numerous presentations over the next 3 days. During and after the meetings, almost all attendees expressed gratitude to the co-chairs, committee members, and the National Academies for enabling this workshop (e.g., "Thank you. This far exceeded expectations. Good to continue this collaboration." [Maj Gen Post, during day 3]).

The committee's process was to look at what is being done now in the Air Force based on current Air Force requirements, to look at what is being done elsewhere, and to compare these, as well as use discussion and committee expertise to identify the areas that can offer further benefit, including items beyond flight crew training. With a few exceptions, the speakers were asked to organize their talks to present what they are doing now, identify the limitations of what they are doing now, identify what they

---

[1] U.S. Air Force, "Link Trainer," Fact Sheet, Posted July 29, 2009, http://www.nationalmuseum.af.mil /factsheets/factsheet.asp?id=3371.

[2] Appendix A provides short biographies of the committee members. The committee reflects extensive expertise in computer science, modeling and simulation, gaming, military operations, and human behavior in stressful environments.

BOX 1-1

Terms of Reference

An ad hoc committee will plan and convene one 3-day public workshop to: (1) examine how simulation is currently used in military services, private industry, and other government agencies, such as the Federal Aviation Administration and NASA; (2) compare alternative uses to current Air Force practices to identify areas where the Air Force can benefit by adopting such practices; (3) examine how current and future technologies will allow the Air Force to gain even more benefit from simulation; and (4) examine how the combination of live training, virtual training in simulators, and constructive/computer generated entities can improve aircrew training. Regarding topics #2-4, the areas where the Air Force can benefit will be grouped into two categories: (1) areas that enhance and/or augment the learning process; and (2) areas that may be used as a substitute for some training requirements with operational systems. The committee will develop the agenda for the workshop, select and invite speakers and discussants and moderate the discussions. The workshop will use a mix of individual presentations, panels, breakout discussions, and question-and-answer sessions to develop an understanding of the relevant issues. Key stakeholders would be identified and invited to participate. One committee-authored workshop report will be prepared in accordance with institutional guidelines.

would like to be able to do, and offer their thoughts on how they can achieve this, particularly in use of technology. The speakers were also asked to frame their presentations in light of needs for simulation expressed by the Air Force using commands.[3] The committee considered all Air Force aircraft types, but fighter aircraft and their missions had the most demanding training requirements and became the main focus of the workshop.

After user needs (requirements) and Air Force supporting activities are addressed in Chapter 1, the remainder of this report is organized around the four numbered items in the TOR, namely, examining how simulation is currently used outside the Air Force (Chapter 2) and how the Air Force might benefit from alternative uses and technologies, especially LVC (Chapter 3). A discussion of (1) areas that enhance and/or augment the learning process and (2) areas that may be used as a substitute for some training requirements with operational systems, as specified in the TOR, is found in Chapter 2 and Chapter 3 as part of the participant dialog. Finally, during the course of the 3-day workshop, common messages, or themes, appeared as a result of various presentations and resulting dialog among the participants. Listed next to each theme are the names of the participants who identified the common message. Details underlying each theme are found in the body of the report. *The report summarizes the views expressed by individual workshop participants. While the committee is responsible for the overall quality and accuracy of the report as a record of what transpired at the workshop, the views contained in the report are not necessarily those of all workshop participants, the committee, or the National Research Council.*

USER NEEDS

During the course of the workshop, three Air Force major commands (MAJCOMs)—AMC, Air Combat Command (ACC), and Air Force Special Operations Command (AFSOC)—presented their needs with respect to LVC training. AMC trains, organizes, and equips the Mobility Air Forces (MAF); ACC does the same for the Combat Air Forces (CAF); and AFSOC's responsibilities for its forces are similar. The abstracts for the MAJCOM presentations are reprinted in Boxes 1-2, 1-3, and 1-4. The leaders of the commands and their staffs, committee members, and many guests spent much time over the 3 days

[3] Appendix B provides a list of workshop speakers and the topics that were addressed during the 3-day workshop.

BOX 1-2

Air Mobility Command

Lt Gen Brooks Bash, Vice Commander

Air Mobility Command is the lead command for rapid global mobility (RGM) and is responsible for guiding the Mobility Air Forces (MAF) community in concept development and force structure. The MAF optimizes the active duty, Air Reserve Component, and Civil Reserve Air Fleet to achieve a cohesive system for RGM effects. RGM, through three core mission areas—Airlift, Air Refueling, and Aeromedical Evacuation, is the key to maintaining global presence and a timely response capability that is the backbone of expeditionary operations, such as supporting strike operations with air refueling or moving forces from the continental United States directly to points of effect.

Maintaining the proficiency of our aircrew is essential to the successful accomplishment of RGM, but sequestration and budget cuts have put flight training time at risk. AMC is looking for more efficient ways to effectively train our crews and align training requirements with the appropriate device. Beginning in 1992, the command began an extensive upgrade of its simulators. All AMC pilot simulators are now the equivalent of FAA Level C (or better), allowing the use of flight simulators for many training events that were previously performed in the aircraft. Currently, an average of 61 percent of MAF pilot flight training requirements is accomplished in a simulated environment. The training is good, but we can make it better.

AMC is upgrading visual systems, improving fidelity, and networking simulators through Distributed Mission Operations (DMO) to capitalize on the efficiencies of live, virtual, constructive (LVC) training. Through DMO, AMC will be able to connect non-collocated receivers, tankers, and Boom Operators to conduct virtual air refueling. By putting a human in the loop, the suspension of disbelief is greatly enhanced; crewmembers are held accountable to entities outside of the box and must work together for successful mission accomplishment. DMO is used by the MAF for daily, persistent training and AMC is looking to expand that capability.

There are several mission sets where simulation is not optimal and aircraft training flights remain essential. Tactical events, such as assault landings, airdrop, and air refueling are not yet fully replicated. Also, as we have already migrated over 60% of our training to simulation, any further migration gives us concern for our ability to gain experience in the mission management aspects of our global mission such as enroute support, aircrew management, Air Traffic Control, C2, and ground support interaction that are crucial for the development of our aircraft commanders. (See second attachment.) Flight training for Loadmasters, Boom Operators, and Aeromedical Evacuation Crewmembers also represent an opportunity as their flight training devices are not as mature as the pilot simulation devices. Indeed, heretofore Loadmasters and Boom Operators gained training as an outcome of required pilot/AC in aircraft flight training, but as we have decreased pilot flight time these crew positions require increased simulator capacity and fidelity to achieve requisite training.

AMC is keenly interested in garnering an expanded awareness of cutting-edge simulation in the aviation industry; ready to capitalize on synergies that will increase the efficiency and effectiveness of aircrew training system.

discussing these needs, their policy and technical implications, and how they could be satisfied. Understanding these needs fully was essential to progress toward identifying how a range of alternative uses of simulation and a variety of simulation technologies could benefit the Air Force.

SPEAKER COMMENTS RELATED TO AIR FORCE NEEDS FOR LIVE, VIRTUAL, CONSTRUCTIVE TRAINING

> LVC is an opportunity for the MAF; but a necessity for the CAF.
>
> —Lt Gen Brooks Bash, AMC

Although Lt Gen Bash's focus was more on efficiencies, which could be gained by moving additional flying hours to simulators, he did recognize that some LVC simulation for training could be very helpful to prepare those MAF elements needed for actual combat, such as refueling and some airlift

BOX 1-3

Air Combat Command

Maj Gen James Post III, Vice Commander

Air Combat Command is the primary force provider to America's warfighting commands to support global implementation of national security strategy. ACC operates fighter, bomber, reconnaissance, battle-management and electronic-combat aircraft. It also provides command, control, communications and intelligence systems, and conducts global information operations. In order to adequately prepare warfighters for future operations across the Air, Space, and Cyber domains, the Combat Air Force (CAF) needs the capability to train and test in a highly realistic and contested environment. This environment can best be replicated using a combination of LVC assets. Advancements in digital technology are enabling the Air Force and Joint communities to integrate the LVC environment into a holistic and realistic training environment where future generations of warriors can be trained. Current training in the Virtual-Constructive (VC) environment is well advanced, but the CAF has a great deal of work to do to integrate VC entities into the live training environment.

Training Advantages of Combat Air Forces (CAF) LVC Capability

Live training will remain a critical and irreplaceable part of CAF training to ensure the entire "system" (aircrew, aircraft, maintainers, supply chain, support functions) is prepared for war. Aircraft must be flown against live targets, surged regularly and subsequently "broken," to validate what works and what doesn't work. CAF aircrew needs to train in real-world conditions/limitations. Examples include: wingtip vapor trails that give away a stealth aircraft's position, altitude block and training rule limitations, high-G environments, inoperative radars or radar warning receivers, and real-world radio communication interference/confusion.

Future VC training will be a critical enhancement to Live training. Once integrated into the live environment, VC will enable a robust, complex, and more cost-efficient threat environment than could ever be replicated by live assets alone. High-end adversary threat capabilities will be replicated in a secure VC environment that is then integrated with live adversary threats. Live and virtual aircraft will engage Live, Virtual and Constructive threats over a secured training network without divulging their full combat capabilities. Live blue air will be integrated with VC support assets (service, Joint or Coalition) to practice synchronized operations that are difficult to replicate in the live environment alone. CAF assets will virtually practice OPLAN missions against constructive Integrated Air Defense Systems that accurately replicate realistic Enemy Orders of Battle. The result is training in a realistic domain where simulated versus live training is only a matter of physical location of the cockpit, and the stimuli of the physical environment.

LVC Operational Needs/Requirements

The CAF LVC environment will exist to provide "expert level" training to operational warfighters and provide an integrated readiness training environment in which warfighters solve dynamic mission execution problems. Today, CAF VC utilizes Distributed Mission Operations (DMO) to connect multiple simulators at varying locations throughout the world for daily team training scenarios—from unit-level package-sized tactics, to large scale exercises among Service, Joint and Coalition warfighters. Tomorrow, the CAF must inextricably link the "VC" to the "L."

In addition to high fidelity concurrent simulators, the CAF requires access to suitable training ranges, airspaces, and training assets for realistic aircrew training. Because the military's training requirements reflect changing technologies, capabilities, and global threat estimates, the AF must continuously review its training requirements and fund for required changes that keep pace with warfighter requirements.

missions. Lt Gen Bash was also interested in ways to help him know where to best spend the next dollar on training. Maj Gen James Post III, ACC, was emphatic about the need for linking VC to L, which is necessary to prepare CAF for the high-end fight. "The CAF wants to evolve to a high fidelity training

BOX 1-4

Air Force Special Operations Command

Col Steven Breeze, Chief, Operations Training

Air Force Special Operations is the air component for United States Special Operations Command and the second largest of the five components behind United States Army Special Operations Command. AFSOC is organized into 3 Wings, the 1ST, 24th, and 27th Special Operations Wings. We also have 2 Direct Reporting Groups, the 352nd and 353rd, the Air Force Special Operations Air Warfare Center, one Reserve Command Wing…the 919th SOW, and a single gained Air National Guard Wing with the 193rd SOW. In many cases our Air Commandos and weapon systems are not assigned to just a single mission set. We frequently execute missions that span across multiple core mission areas, almost always in conjunction with our Army, Navy or Marine special operations partners. Those mission sets range from specialized air mobility to precision strike to ISR.

This past summer, the new AFSOC Commander refocused and reviewed the Commands priorities and highlighted the need to improve our training. Out of those extensive reviews, the Command deemed the importance of transforming our training to optimize human performance. Multiple lines of effort were developed to improve our training and refocus standards on excellence. To reach those standards, our goal is to leverage the synthetic environment and state-of-the-art training methods. While all of our simulators are now the equivalent of FAA Level C or better, we do not have simulators collocated with each operational squadron. We are "late to need" programming simulators for our next generation aircraft. While our training systems are not broken, we need to take advantage of the synthetic environment to eliminate the obsolescence of our training systems.

As part of our training transformation, we have systematically reviewed all currency requirements in all of our MWS's refocusing continuation training to include the simulator. We determined multiple events can be better trained or more safely trained in the simulator. Through this process, we hope to free up aircraft time to increase the amount of joint training we can conduct with our partners and provide more combat power downrange. While we have not reduced the flying hour program, we are setting conditions to absorb a future decline.

Due to our diverse mission sets and the importance we place on crew resource management, there are several areas where simulation is not optimal. While we have not reached the max amount of simulator events capable of being logged in the simulator, we are quickly reaching the limit due to several factors. (1) While the visual systems in our simulators are excellent, they are showing their age (8-10 years) and therefore we cannot replicate the full tactical environment. (2) AFSOC rapidly upgrades aircraft; simulator programs and funding are frequently left behind (late or unfunded). (3) Most of our MWS's heavily incorporate the "crew concept"; however the simulators and fuselage trainers or back-ends are not linked. (4) The aero models in some of our simulators rely on engineering data and not flight data limiting flight fidelity. (5) Complex databases include six or more layers (imagery, elevation, material, features, light, 3-D models, and radar) and are extremely time consuming and expensive to build manually.

AFSOC is still in the infancy stage taking advantage of Distributive Mission Operations (DMO). Currently, each crewmember participates in one DMO event per semi-annual period. Challenges remain leveraging the capabilities of networked simulation efforts. We have a lack of manpower and simulator capacity to ensure every crew in AFSOC is capable of training in the DMO environment. Also, our threats are not validated or centrally monitored to ensure fidelity. Finally, there is no standardized Multi-layer Security Solution to enable training with 5th generation fighter aircraft.

environment through integration of dynamic L, V, and C."[4] Maj Gen Post was adamant about not cutting live flying hours: "VC is outpacing L . . . but L is a necessity for the CAF. We need to focus on the 'dash' between L-VC so we can connect the VC to L." Col Nathan Hill, Chief of ACC Flight Operations, then added several comments. Col Hill stated that realistic training is a requirement for the CAF to ensure that

[4] The level of simulation fidelity required for training tasks is a topic that recurred during the workshop. The discussion would often refer to the need to understand the level of simulation fidelity required for training effectiveness. The value in doing this was to avoid the cost and technical risk associated with developing a greater level of fidelity than necessary for training effectiveness for a particular mission. The importance of ensuring correct "muscle memory" for controlling the vehicle through training in addition to higher-level decision making was also emphasized during discussions.

the Air Force is prepared for *all* contingencies across the range of military operations. In addition, Col Hill believed that the desired end state for CAF is full LVC: putting virtual and constructive into live aircraft. He further noted that CAF needs to determine the right balance of live fly and simulation (the equation will likely be changed every 1-3 years) and needs to resolve security concerns as we put more and more onto various networks (an ongoing concern). Finally, Col Hill stated that CAF also needs technology advances to ensure full LVC (e.g., What waveform will live aircraft use? and How will the VC be put into each type of aircraft?). Many participants pointed out that AFSOC's requirements fell between AMC and ACC with respect to LVC.[5]

COMMITTEE COMMENTS RELATED TO AIR FORCE NEEDS FOR LIVE, VIRTUAL, CONSTRUCTIVE TRAINING

Robert Allardice, former vice commander of AMC, noted that complexity and advances in warfare have moved to the point where legacy training platforms are inadequate in producing operationally ready aircrew. Therefore, according to Mr. Allardice, the Air Force must undertake LVC training methods to integrate 5th-generation aircraft [red and blue] into its "simulation" training portfolio because the current construct is inadequate. "Operationalize" LVC and have acquisition programs address that. Recent advances in technology allow for investments in distributed training with a very favorable return on investment (due to cost avoidance). Mr. Allardice submitted that this is the efficiency side of the argument that seems to be the focus of AMC. Advances in simulation must have the following common attributes: concurrent, dynamic, realistic, and degraded operations.

John Corley, former ACC commander and former Air Force vice chief of staff, noted that the Air Force needs both a more effective and efficient approach for the training environment. He went on to say that ACC's demands tend more toward the effectiveness imperative while AMC sees the greatest benefit (while not exclusively) in efficiency, especially given the severity of fiscal constraint. Mr. Corley offered that both commands can benefit from the development of a realistic training domain where simulated versus live training is only a matter of physical location of the entity and the stimuli of the physical environment; an approach that potentially yields this realistic domain is through a properly constructed LVC capability. Finally, he noted that development of the above can include a process to demand compliance with requirements and funds for required changes that will keep pace with warfighter requirements.

Steve Detro, a business development lead for Lockheed Martin Mission Systems and Training, noted that, since 1986, MAF, and AMC specifically, has operated under the policy of using FAA Level C and FAA Level D equivalent flight simulators to train for 100 percent of transport aircrew certification. He went on to say that this policy has generated tremendous savings and continues to do so due to the fidelity of the aircrew produced. Mr. Detro believes that some elements of aircrew experience development have been identified as needing additional focus of training (e.g., airmanship, judgment development, and overall seasoning of aircrew) and would benefit from a higher level of virtual environment fidelity in simulation. LVC could provide more efficiencies and cost savings for high-risk mission training tasks. Finally, Mr. Detro noted that LVC could provide higher-level skill development, such as "edge of the envelope" training for missions like air refueling, air assault, airdrop, etc. Pertaining to CAF, and ACC specifically, Mr. Detro stated that since the development of Distributed Mission Operations (DMO) networked simulators in the early 1990s, ACC has fielded and is using simulation to do team training between disparate air platforms in progressively more complex operational environments—for example, training for multi-ship tactical, joint service operations, coalition exercises,

[5] There were several comments from participants on the importance of LVC in training/mission rehearsal for integrated Strike packages. It was mentioned that individual components of a package could be trained on simulators, but combined packages were trained using actual aircraft, which is expensive and risky. Effective use of LVC to train combined packages for ACC is critical.

and large force exercise work-up (i.e., Virtual Flag and Red Flag). He noted that ACC has moved to using high-fidelity simulation for a larger percentage of its training versus live fly, but does so under the philosophy of using the simulation sortie to complement the quality of the live fly sortie. The ratio of simulation to live fly is different for each aircraft type, due in part to the different levels of fidelity of each simulator. The newest fighter flight simulation technologies are enabling the F-35 pilot training center to move more than 50 percent of flight training sorties out of aircraft live fly into the virtual reality flight simulator. Other fighter training programs are also being enabled, through simulation fidelity improvements, to move a portion of their training sorties to virtual simulation as the fidelity of each aircraft simulator permits. Mr. Detro observed that LVC is an imperative for both 4th- and 5th-generation fighter operations, a must-have to complement current levels of live flight operations. DMO, the predecessor technology to LVC, currently supports approximately 25 percent of the high-end training and tactics training in the Air Force. Mr. Detro believes that sustained funding is required to fully realize the benefits.

AFSOC, Mr. Detro noted, uses distributed (networked) simulation for a very large percentage of its crew training due to high dependence on total crew proficiency in high-tasking mission scenarios; there is 100 percent linking of simulators across AFSOC. Further, Mr. Detro observed, AFSOC requires all crews to use simulation for 30-40 percent of all training. AFSOC, as he noted, has the near-term goal of fielding flight simulators at all operating bases to be utilized for continuation training and continued mastery of high-fidelity aircraft equipment (e.g., night-vision goggles, electronic warfare, and terrain following radar, weapons, sensors, communications, and navigation systems).

Ray Johns, former AMC commander, noted that the strategic environment has changed—we are not at war, so there is no choice but to put red missions in some kind of virtual environment. Harry Robinson, SimLEARN National Program Manager at the Veterans Health Administration, offered that the demands of 5th-generation aircraft do not afford a full spectrum of training for aircrew in a live simulation domain. Mr. Robinson went on to say that use of simulation is critical to ensuring that warfighters are ready on day 1 of combat operations; there are little resources, time, or tolerance to support learning during battle. Mr. Robinson added that there are significant differences between training for currency (based on periodicity) and proficiency (based on competency); just because a pilot drops a bomb once every 3 months, it does not mean that pilot can hit the target. Determining the amount of funding for training based on periodicity is a much easier problem to solve than proficiency. Mr. Robinson submitted that some training is accomplished during actual mission performance (e.g., combat missions, search and rescue, command and control). This training addresses both competency and currency.

Michael Zyda, director of the Game Pike Laboratory at the University of Southern California, believes that the Air Force cannot turn on the secret equipment in training without giving away the secrets. He noted that network security causes training problems, mostly because multiple networks are connected, and he said that the intranets are fine with respect to security. He also indicated that National Security Agency (NSA)-certified multilevel security is needed. Mr. Zyda noted that there are hard-coded requirements in the contracts; consequently, emerging behaviors are not modeled. How to make the environment more dynamic is an issue, in his opinion. Reliability is so high in planes today that they only see systems failures in the simulators. Mr. Zyda submitted that there appear to be assumptions that there will always be a "man-in-the-loop"; he believes the future is clearly autonomous systems. Finally, Mr. Zyda offered that AFSOC wants synthetic environments and state-of-the-art training devices; AFSOC has special mission equipment that must be in the simulator. For continuation training, the desire is to do all of it in the simulator. He noted that AFSOC would also like higher-end events in the simulator, but they are not there yet. The dialog about user needs led to the first key theme of the workshop.

> **Theme 1.** For current and future warfighters to be operationally ready on a continuous basis, realistic training in a simulated environment is critical. For Air Combat Command, in particular, training in the "live" (L) construct linked to "virtual constructive" (VC) is imperative for mission success. For Air Mobility Command training, VC is critical, but its requirements are somewhat

fewer with regard to linking to the L environment. With respect to LVC training, Air Force Special Operations Command's requirements are between Air Combat Command and Air Mobility Command (Ray Johns, John Corley).

Further exposition of user needs was offered by Steve Detro:

- For AMC: (1) additional simulation technologies to expand the number and realism of real world experiences for aircrew (i.e., air traffic control congestive environments, mission management, crew resource management, crew fatigue); (2) training technologies that accommodate the different learning styles of today's pilots; (3) methods to objectively measure aircrew competency (note: mission essential competencies and pilot evaluation techniques that were developed at the Warfighter Readiness Research Division of the 711 Human Performance Wing, Human Effectiveness Directorate, Air Force Research Laboratory [711 HPW/RHA] by Dr. Wink Bennett); (4) use of the "science of learning" to optimize the training delivery methods and more efficiently utilize the full range of fidelity levels provided by a family of simulators; and (5) affordability.
- For ACC: (1) a more efficient way to develop, integrate, and deliver a persistent, cost-effective, LVC network across multilevel security simulators; (2) concurrent simulators that more accurately replicate the most current aircraft capabilities; (3) higher-fidelity simulators that accurately replicate aircraft systems, engines, avionics, aerodynamics, weapons systems, sensors, environments, threats, and communication systems; (4) flexibility in the simulation that enables the accurate modeling of combat conditions, to accurately simulate the unpredictable nature of operations in the environment of contested and degraded operations; (5) more efficient process for cross-domain network security; and (6) validated threat systems that are physics-based and exhibit intelligent behaviors.
- For AFSOC: (1) accurate validation of the optimal ratio and training balance of simulation "virtual" training versus aircraft-based "live" training; (2) upgrade of AFSOC's legacy simulators to fix limitations (i.e., fidelity of visual environments for night-vison goggles at low-level operations, aero models, concurrency, faster scenario development, and physics-based electronic warfare models); (3) simulation of ramp operations to reduce the number of vehicle-aircraft collisions; and (4) better implementation of the ability to generate simulation scenarios that present situations or events that surprise aircrew during simulation evaluations.

Relatedly, John Corley offered that chasing physical fidelity may be a fool's errand. "Sufficient fidelity" could be delivered through "perception of reality." In turn, Mr. Corley submitted, we could achieve desired and measurable behavior. Steve Detro suggested that the Air Force continue to analyze potential benefits of virtual reality and gather measurable data to substantiate that the higher the fidelity, the higher the benefit. Finally, Harry Robinson noted that realistic simulation and credible simulation are not interchangeable terms. Realistic simulation is the measurement of fidelity and resolution. Credible simulation is the measure of trust in the simulation for providing an immersive training environment that supports the suspension of disbelief.

AIR FORCE PROGRAMS TO SUPPORT USER NEEDS

Representatives from Air Force Headquarters described broad, top-level guidance regarding simulation that reaches all major commands and nearly all core functions of the Air Force (see the abstract in Box 1-5). Below is a relevant extract from one piece of this guidance. Figure 1-1 depicts a notional end-state for LVC-Operational Training (LVC-OT).

> This LVC-OT Flight Plan highlights the areas and item that need particular attention to advance the LVC-OT program and realize its full potential. The specified enabling processes address a governance structure, processes, and infrastructure—all essential to furthering LVC-OT

BOX 1-5

Headquarters Air Force and Air Force Agency for Modeling and Simulation

Brig Gen Eric Overturf, Mobilization Assistant to the Director of Operations, Deputy Chief of Staff for Operations, Plans and Requirements

The Air Force Agency for Modeling and Simulation (AFAMS) through the Headquarters Air Force A3 is the lead agent for centralized management of Air Force cross-functional and shared live, virtual, and constructive (LVC) foundational capabilities and resources supporting Air Force Service Core Functions. The AFAMS mission is to provide seamless integration of cross-functional LVC environments for operational training that allow warfighters to maximize performance and decision making. AFAMS serves as the HAF lead for Air Force LVC foundations and integration with the Department of Defense, Service Components, other government agencies, international partners, academia, and industry. This mission provides the necessary development and implementation of standards for common access and interoperability within the LVC domain for efficient and secure global operations (AFMD56 14 JANUARY 2014).

This summer, the Secretary of the Air Force and the Chief of Staff of the Air Force unveiled the Air Force's 30-year strategic vision and introduced the concept of "strategic agility" and stated, "One of the more promising paths to agility in operational training and readiness is in the area of Live-Virtual-Constructive training." The Air Force is in a period of training transition due to available emerging and advanced technologies, fiscal constraints, and inability to train to the actual capabilities of our latest weapons systems highlighting the need to transition from the historical focus on live training to achieve warfighter readiness. There will be challenges at the forefront of this transition, but these challenges are not insurmountable. These challenges do merit closer collaboration with our sister services and our industry partners. Air Force (and national) readiness increasingly depends on the ability to harness and manage complex training systems and systems of systems. To summarize, "Readiness through LVC" is based on Strategic Guidance, OPLANs, and CCDR requirements/demands, utilizing the capability and capacity of manpower and resources on a timeline that is balanced by "fight tonight versus modernize for tomorrow."

The programs encompassed within and touched by the LVC capability are numerous; they reside in every MAJCOM and nearly all 13 Air Force Core Functions. This is an important point because the MAJCOMs remain the key force providers who organize, train, and equip; Headquarters Air Force provides the overarching and broad strategic guidance ensuring standards and standards development are a foundation to the future of LVC. Headquarters Air Force A3 wears two hats in the planning/programming world: (1) as the lead and direct input source for LVC Foundational requirements and (2) as the programming advocate for operational training to help shepherd and support the MAJCOMs/CFLs issues through the Air Force Corporate Structure. Headquarters Air Force conducts support/advocacy/engagement in accordance with the SECAF/CSAF LVC Flight Plan signed in February 2013 and are working to codify this process in enduring and binding documents such as AFPD 16-10 Modeling and Simulation, AFI 11-202V1 Aircrew Training, 11-2MDS-V1 MDS Training, AFI 36-2251 Management of Air Force Training Systems, and AF Mission Directive 56 Air Force Agency for Modeling and Simulation, to name just a few. Our top priorities are to (1) support and advocate on behalf of the force and codify LVC standards and (2) provide support and Authorizing Official duties for Cybersecurity and Authority to Operate/Connect for LVC-related training systems.

The Air Staff under HAF/A3, Gen Field, developed these four enduring lines of effort to capture the LVC strategic focus: (1) LVC Foundations—develop policy and guidance that enable effective, efficient, training, test, and analyses in a secure LVC domain; (2) Aircrew Training Devices (Sims)—develop Air Force strategy and policies that align with COCOM requirements and Joint policy that provide affordable ATDs with timely concurrency, sufficient fidelity, and appropriate connectivity; (3) Distributed Training—develop the appropriate Air Force strategy and policies to enable effective, secure, distributed training in Air Force and Joint synthetic training environments; and (4) Full LVC—develop an Air Force strategy that aligns with Joint programs to integrate live aircraft, space, and cyber systems with virtual battle spaces.

Ultimately, the goal is a fully integrated operational training continuum, where "live" aircraft on a range fully integrate with "virtual" participants in simulators and "constructive" entities representing Red/Blue Air, Threats, Ground Forces, and Targets, all supported through readiness/distributed training centers and range control complexes for full spectrum combat ops training.

NOTE: The Air Force provided the following document to the workshop participants to illustrate current initiatives related to LVC training: "Bullet Background Paper on Air Force Live, Virtual, Constructive Vision and Strategy," Col Crites/AFAFMS/CC/970-5701/srfs/18 Nov 2014; Air Force LVC-OT Standards Profile.

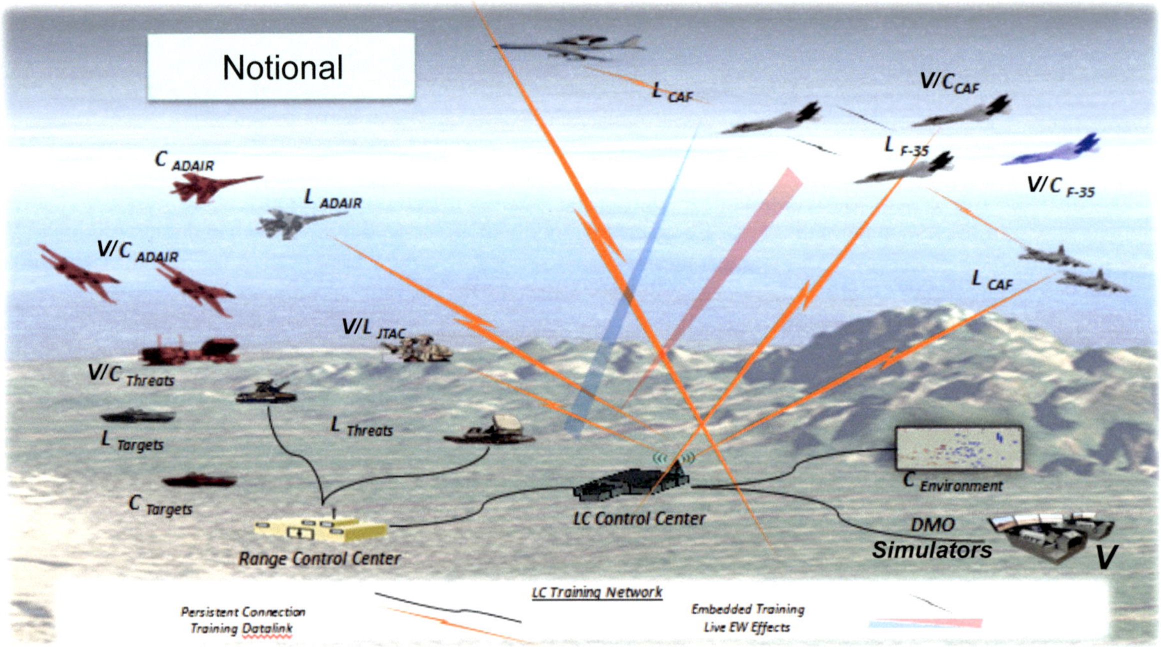

FIGURE 1-1 Notional end state for live, virtual, constructive-operational training (LVC-OT). NOTE: The "nirvana" end state for Air Force simulation is where all Air Force weapon platforms are linked together to enable realistic, distributed mission operations in a live, virtual, constructive environment. SOURCE: Brig Gen Eric Overturf, Mobilization Assistant to the Director of Operations, Deputy Chief of Staff for Operations, Plans and Requirements, Headquarters U.S. Air Force. SAF/PA Approved for Public Release 2014-0569.

> capabilities. Four key focus areas (LVC foundations, weapon system simulators, distributed training, and full LVC) are introduced and will lend permanency and stability to current LVC activity. Air Force-level requirements and investment strategies are also established to ensure operational and technical priorities are addressed, funded, sustained, and are in-line with operator readiness requirements. Finally, the LVC-OT Flight Plan identifies roles and responsibilities at all levels within the Air Force and provides a time horizon for specified actions.[6]

Several committee members reacted to the issues of attention to LVC at top levels of the Air Force and broad Air Force application of simulation technologies. First, John Corley noted that consensus must be reached on the current vision (modified at appropriate frequency) for LVC and that there must be an advocate, with both responsibility and authority, to deliver vision, strategy, and strategic plan for LVC. Mr. Corley submitted that Air Force communities (i.e., MAJCOMs) have arrived at the limits of live training. Further, he believes that separate and distinct virtual (simulation) or constructive approaches, when applied in an additive fashion, will not meet the knowledge transfer threshold today, much less the future. The appropriate integration of L, V, and C can achieve the full spectrum of needed training while also benefiting those requiring training across the full range of military operations. Mr. Corley offered that simulations growth through a prudent, commonly accepted LVC approach can provide increased learning benefit for the full complement of mission capabilities and developmental activities.

On a related topic, Ray Johns, committee co-chair, stated, "The Air Force needs an overall LVC strategy. The Air Force needs to state the LVC requirements, which will drive an acquisition strategy, which will drive a program." Committee member Richard Reynolds, former vice commander Air Force Materiel Command, noted that establishing a durable understanding of LVC's relative worth compared to

---

[6] U.S. Air Force, *United States Air Force Live Virtual Constructive Operational Training Flight Plan,* February 22, Washington, D.C.: Headquarters U.S. Air Force, 2013.

other components of readiness is necessary. LVC is ultimately going to have to compete against other Air Force programs and priorities. Of course, he offered, work will be needed to define "durable" and "relative worth," and, when done, one result will be discarding things that are not necessary. Strategic communications (aka "marketing") will be important. In the eyes of Harry Robinson, committee member, there would appear to be many more applications for employment of simulation in Air Force training environments than has been addressed. With major emphasis on ACC, AMC, and AFSOC, Mr. Robinson thinks there would be value in opening the aperture for a bigger simulation umbrella to include Air Education and Training Command, Air Force Global Strike Command, Information Dominance, Air Force Space Command, and Air Force Research Laboratory. Mr. Robinson believes there is also a need to have simulation solutions that are driven to support inter-service training events. Committee member Michael Zyda offered that, clearly, the Air Force could use a chief architect and standards for its LVC systems. That is one of the biggest messages. The dialog about top-level guidance led to a second key theme of the workshop.

> **Theme 2.** Establishing stated requirements for live, virtual, and constructive training as well as implementing a live, virtual, constructive training strategy and governance model could greatly benefit the Air Force across its full range of missions. This undertaking will likely mean establishing a durable understanding of live, virtual, and constructive training's relative worth compared to other components of readiness (Ray Johns, Donald Fraser).

Speakers from the Air Force Life Cycle Management Center and the Air Force Research Laboratory described various research and development, acquisition, and sustainment efforts under way to satisfy the top-level guidance and meet the user needs (Boxes 1-6 and 1-7). This part of the workshop delved into more technical detail. Illustrative comments from committee members appear below; some of these comments feed back to the needs addressed earlier, while others are precursors to more broad-based comments, which arose later in connection with discussions of a different approach for implementing a simulation architecture.

Committee member Robert Allardice noted that the Air Force simulation roadmap appears to be very immature (standards, disciplined investment, adaptability, distribution architecture, etc.) and that there seems to be a role for a "simulation" integrator across all platforms. Committee member John-Paul Clarke, associate professor in the Daniel Guggenheim School of Aerospace Engineering at the Georgia Institute of Technology, submitted that the flexibility and fidelity that is desired by the Air Force stakeholders will require a modular simulation framework where all possible (or at least a large number of) combinations of L, V, and C elements can be put together so that individual units can schedule and control the conduct of complex or high-end training. He believes that such a framework will be expensive; thus, the development plans must include a transition plan that is dynamic and can respond to variances in funding to ensure that new capabilities are provided at the end of any fiscal year. Dr. Clarke believes time synchronization will be a challenge and that predictive cueing is an obvious approach to mitigating the effects of latency. Another possible approach, according to Dr. Clarke, could be to mix event-based and time-based simulation such that event messages are sent in parallel to real-time data exchange to ensure that specific things that must happen at a certain time actually do occur at that time. Also, for agent-based simulation, Dr. Clarke believes that one needs to know which agents are involved, how much they are involved, and, especially for VC into L, who is the training target.

Committee member Pamela Drew, executive vice president and president of Information Systems, Exelis, Inc., provided that LVC technology has advanced over the past 15 years or so in industry and laboratories, and solutions to some of the Air Force gaps do exist (e.g., the need for virtual reality in heads-up displays). However, she noted, there are major gaps in terms of operational needs of the Air Force to apply LVC to their mission set in a practical way. These gaps include ways to address safety, security (particularly in external DMO networks and coalition efforts), and standards for weapon system interface modifications to achieve interoperability and integration. (Note: this can be referred to as "Operational LVC.") Ray Johns submitted that the Air Force has a need for mission-oriented investments

BOX 1-6

Air Force Life Cycle Management Center Simulators Division

Col Daniel Marticello, Chief

The Simulators Division is the U.S. Air Force's primary agent for the acquisition, sustainment, and modification of aircraft training systems, including flight simulators, maintenance training devices, simulator interoperability solutions, and related services. The division is a component of the Air Force Program Executive Office Agile Combat Support Directorate, located within the Air Force Life Cycle Management Center, Air Force Materiel Command, Wright-Patterson AFB, Ohio.

The Simulators Division consists of over 400 acquisition professionals representing the program management, engineering, contracting, finance, and logistics management fields working together to provide solutions to a variety of ACC, AMC, AETC, AFSOC, AFGSC, and foreign military training requirements. We are responsible for over 40 aircrew and maintenance training system programs, executing a more than $1 billion annual budget for systems and services at over 100 locations worldwide. In addition to aircraft simulators and training devices, the division manages the Air Combat and Air Mobility Distributed Mission Operations programs, providing the Air Force's only live, virtual, and constructive (LVC) operational training capability.

The leadership of the Simulators Division is focused on seizing opportunities for innovation within the sphere of training, cost-capability trades, and the state of the simulator industry. A large modification to an existing weapon system or the procurement of a new one to perform an existing mission presents an opportunity to scrub how training is provided. Simulator technology has moved significantly forward over the past decade in the areas of fidelity and networking. Training system methodologies have also matured, especially within the private sector, allowing more training objectives to be "off-loaded" and "downloaded" to simulators and accomplished at a lower cost.

The way forward to ensure that we capture these advances in capability and the promise of lower cost is to first conduct a Training System Requirements Analysis (TSRA). This study effort looks to capture all of the required learning objectives, throughput and availability expectations, and technology available. This information can then be used to support industry proposals on how best to deliver the training and what simulator devices are proposed. This approach allows industry to bring innovative solutions to the table in a best-value, trade-off type of competition. Subsequent CLS/TSSC and modifications are delivered via a separately competed contract vehicle following an initial period of interim contractor support provided under the production contract. TSRAs are also essential in understanding where best to apply the power of the LVC construct. An informed view of what objectives require interaction between the L, V, and C aspects of training will allow the Air Force to best apply limited resources.

Balanced cost-capability trade-offs are essential in this time of shrinking budgets. The Simulators Division is committed to utilizing data from existing contracts to close the feedback loop. Capability provided should match the level of capability needed. Reductions in capability should also be considered if a large savings can be obtained without a negative effect on mission accomplishment.

The simulator industry is experiencing a change in environment. Small Business Set-Asides, LPTA source selections for sustainment, and data rights are all areas that have unintended consequences. It is wise to understand the macro-level implications of decisions made at the individual program level.

to support LVC to prioritize where to put the next dollar (see Figure 1-2). The most challenging mission and biggest gap, according to Mr. Johns, is the mission set of training for the peer/near-peer adversary against 5th-generation systems; this is the integrated capability for which the Air Force must have LVC at a level that does not exist today. Without it, in the opinion of Dr. Drew, it is very likely that the Air Force is not going to be adequately trained for all threats. The MAF mission can benefit from such an LVC capability in terms of mission support, but will also reap higher dividends in efficiency (i.e., savings). In addition, Dr. Drew noted that the MAF (by repurposing the efficiency savings) could train for higher-end capability. The core architecture for such LVC exists, she believes, and it needs to be assessed for scale, robustness, and extensibility, among other things, as well as for what is needed to implement the Operational LVC to support the 5th-generation scenario. Lastly, Dr. Drew believes if that can be solved, the rest of the missions will be a subset of the solution.

BOX 1-7

Air Force Research Laboratory

Winston Bennett, Division Technical Advisor
for Training and Assessment Research, 711 Human Performance Wing

The Warfighter Readiness Research Division of the 711 Human Performance Wing, Human Effectiveness Directorate, Air Force Research Laboratory (711 HPW/RHA), is the Air Force's premier research and development organization for education and training. The division pioneered the development of Distributed Mission Operations in Collaboration with Air Combat Command. The division has also led the development of methods and tools to persistently gather and track mission performance and proficiency data for the development of more targeted approaches to training. The division and its operational, industry, and academic partners continue advancing the state of the art in learning, performance, and modeling theory and practice. The division also continues to pioneer and advance distributed mission training and live, virtual, and constructive training methods and capabilities and our research continues to drive the Air Force's vision and investment for the future of operational readiness training.

Recent Highlights and Advances

The division is growing our involvement in the Human Systems Community of Interest, promoting stronger collaborations with industry and our international collaborators. Further, the division is creating and transitioning proof of concept developments in learning management and performance measurement technologies, game-based applications for maintenance training, unmanned aircraft operations, low-cost 5th-generation tactical training, and agent development for autonomous operations, man-machine teaming, and increasing the realism and credibility of live, virtual, and constructive training environments. Our groundbreaking research in cognitive models and agents continues to define and push the science and practice state-of-the-art with successes like the synthetic teammate validation work, the growing collaboration with the American Heart Association and the Defense Health community, and prototypic agent-enhanced sensing for autonomous operations. The team is making great progress in integrating agents into operational training simulations to both improve the credibility of the environments for training and also to increase their efficiency by reducing the need for human "white force" support through the use of model-based agents and avatars. Finally, we completed our first distributed live, virtual, and constructive Close Air Support training trial with the U.S. Army, and we also completed our first and very successful demonstrations of medical operations training research technologies for critical care air transport teams, emergency responders, and pararescue personnel.

Looking to the Future

Of course, we are also mindful of the need to continue to look to the future and to ensure that the developments we make today are meeting the operational demand signals we have and are foundational to continued advancements down the road. Our current work has a strong emphasis on helping the Air Force realize its vision for realistic and secure live, virtual, and constructive training, but it is also a pointer to our future directions for personalized, performance-based learning and readiness assessment. In the future, our education and training systems must be agile and responsive to create the resilient Air Force workforce for the future fight that is more responsive, realistic, and pervasive than we know our adversaries will be.

In the eyes of Harry Robinson, current and planned capabilities can meet Air Force needs. The big challenge is drawing lines to define acceptable capability levels at a given point of time that will be acceptable to meet Air Force training requirements. Otherwise, Mr. Robinson notes, it becomes a "death spiral" development. By definition, Mr. Robinson noted, all models are wrong or incomplete; however, some models are useful. He believes it is unreasonable to recreate the actual world in a virtual environment; the challenge is met by acquisition of sufficient simulation to meet the requirement, not more. Mr. Robinson thinks that simulation-based training can and should be focused on specific flight regimes. Analysis of mission conduct, he said, should include disaggregation of specific tasks—from

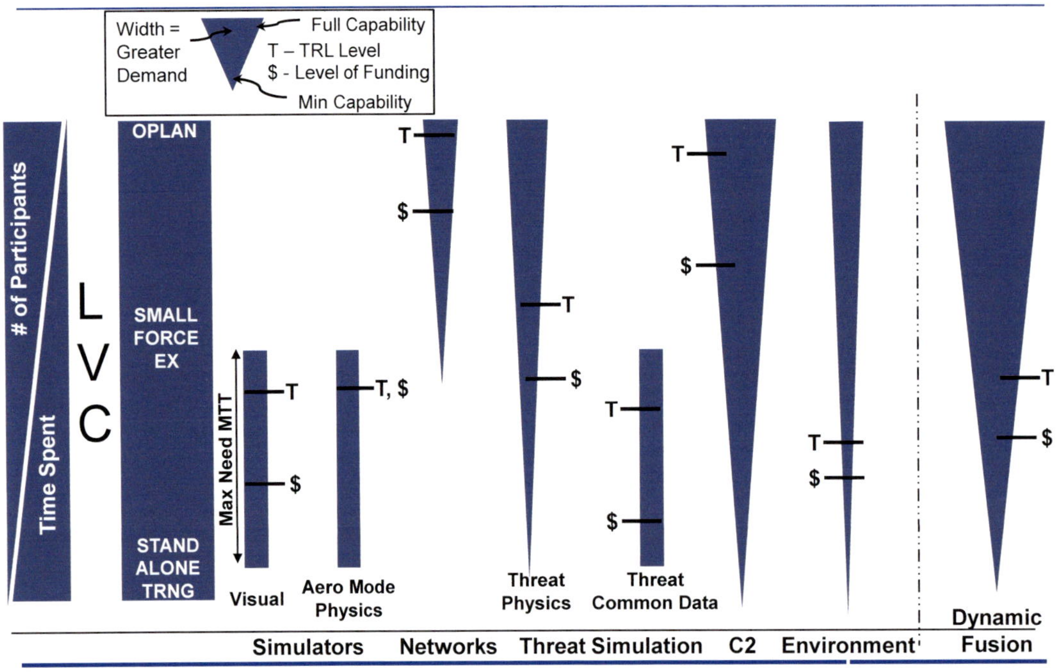

FIGURE 1-2 Notional simulation assessment methodology. SOURCE: Ray Johns.

mission brief to man-up, launch, conduct, land, and debrief. In addition to end-to-end training, Mr. Robinson said, task-trainers and games present unique opportunities to maximize training resulting in proficiency improvements. Finally, he noted that cyber and security demands require attention in modeling and simulation for the training domain; these cannot be effectively backward-engineered into the solution.

2

How Simulation is Currently Used by Military, Industry, and Government Agencies

OTHER MILITARY USERS

Maynard Zettler, director of research and engineering, Naval Air Warfare Center Training Systems Division, discussed the Navy's simulation activities (Box 2-1). The important focus was on the Naval Aviation Enterprise's (NAE's) initiative to improve training by optimizing live, virtual, and constructive (LVC) simulation to match the Navy's recent thrust of integrating its warfighting capability across mission areas, platforms, sensors, weapons, and kill chains. Mr. Zettler explained that this new integration concept differs from prior "stovepipe" approaches and has support at top levels of the Navy.

INDUSTRY USERS

Speakers from Lockheed Martin and Boeing presented simulation approaches of the large U.S. aircraft manufacturing industry (Boxes 2-2 and 2-3), whereas CAE, Inc., and FlightSafety International presented approaches of smaller but important simulation entities (Boxes 2-4 and 2-5). These presentations covered a range of simulation activities and concepts, including small, head-mounted visual displays; large and complex simulators; pilot training and training for other skills (e.g., maintenance); architectures having "the world" embedded in an individual platform simulator; and "world-centric" architecture from which individual platform simulators extract common data (e.g., weather, terrain).

GOVERNMENT AGENCIES AND OTHER USERS

Jeffery Schroeder, chief scientific and technical advisor, Federal Aviation Administration (FAA), and Bimal Aponso, chief, Aerospace Simulation Research and Development Branch, National Aeronautics and Space Administration (NASA), presented the simulation approaches from these two large agencies of the U.S. government (Boxes 2-6 and 2-7). Mr. Aponso offered that NASA has a substantial aeronautical simulation capability, which can be made available at cost to outside users (e.g., for simulating aspects of national airspace) but is no longer central to the agency's main mission of space. Mr. Aponso noted the difficulty of retaining relevant skills in aeronautical simulation activities at NASA. Finally, he discussed NASA's development and testing of an LVC architecture for researching integration of unmanned aerial systems into the National Airspace System (NAS). As part of this development, NASA is characterizing latencies throughout the LVC using a realistic NAS air-traffic simulation and is developing improved communication protocols to integrate the L with the VC components. Mr. Aponso offered that this work may be useful to the Department of Defense (DoD).

In contrast, Dr. Schroeder explained that the FAA oversees U.S. civil aviation in which pilot training and checking is done predominantly in simulators, and the agency sees no reason to change this paradigm. Dr. Schroeder's video clip at the workshop showing pilot reactions to the introduction of highly unusual events into simulator routines was of special interest.

BOX 2-1

Naval Aviation Enterprise

Maynard E. Zettler, Director–Research & Engineering, NAWCTSD

The Naval Aviation Enterprise (NAE) is undertaking multiple initiatives to improve training optimization and proficiency. Central to many of those initiatives is the utilization and integration of Live/Virtual/Constructive (LVC) simulation to augment and improve training. The "LVC in Naval Aviation Training" presentation will focus on the NAE's operational context and integration across the LVC domains. The ultimate objective is the optimized use of LVC to improve the NAE's Integrated Warfighting Capability across mission areas, platforms, sensors, weapons and kill chains. The presentation will address the LVC Training Requirements Path and the process for defining not only what needs to be trained but also utilizing the science of learning to understand the most effective methods to accomplish the training and sustain the requisite skills. The challenge is not just within a given platform but across platforms and the complementing entities in the kill chain(s). Representative examples of current initiatives will be provided, coupled with a discussion on investment gaps and barriers to success.

BOX 2-2

Lockheed Martin Mission Systems and Training

Rick Boggs, Senior Fellow

Lockheed Martin and the U.S. Air Force ATARS II program have engaged in successful human performance engineering. For the past four years there has been an activity centered around Training Transformation that has made some very good progress. With the entry of the F35 into the fleet comes a challenge of Live Virtual Constructive environments. Lockheed Martin is working on a LVC environment known as ACES to address the inclusion of 5th-generation aircraft. Today's training requirements require a 360 degree visual display that is expensive to purchase and operate. I think the requirement should be adjusted to allow for the new man-wearable technologies. These new technologies save considerable expenses and do not reduce the quality of the visual display to the air crew.

The last speakers of the workshop were from the University of Toledo and the State University of New York at Binghamton, and they presented simulation approaches in the medical field (Box 2-8) and an academic modeling approach (Box 2-9), respectively.

FOLLOW-ON REMARKS

After commenting at the meeting, Sharon Conwell, senior research psychologist, Warfighter Training Systems and Performance Assessment Branch (RHAS), Air Force Research Laboratory, made a special effort to provide written comments regarding the medical presentation.

> Thank-you for allowing AFRL/RHAS (Wink Bennett and I) to attend the LVC AFSB study discussion. I found the meeting most informative. You requested that I provide a sentence or two for your study regarding my comment about the 88th Med Group at WPAFB (Wright-Patterson Air Force Base). According to cost research done by Mr. Jacob Arnst at the 88 MDSS/SGSRM and reported by Col Penelope Gorsuch, Deputy Commander of the 88th Medical Group (88MDG/CD), when comparing the 88th medical group to a comparable private sector medical facility, the medical group loses 38 cents on every dollar. Some portion of the 38 cents is more than likely related to training/readiness costs. Every hospital has significant training costs, but military treatment facilities have additional readiness training costs above those of a private sector hospital. The researchers at

BOX 2-3

The Boeing Company

Steve Monson, Chief Architect, Technical Fellow-Simulation and Training

We are in most certainly in a "do more for less" environment, with a need to provide more effective training for reduced costs. Leveraging innovations in commercial technologies and other industry investments like Integrated—Live Virtual Constructive (I-LVC) simulation require a partnership with the Air Force to maximize utility and benefits. Industry is well equipped to research, develop, and tailor technologies for training, and the Air Force is equipped to evaluate and transition these technologies to acquisition programs.

Leveraging Commercial Technologies

Low-cost commercial immersive visualization technologies such as Oculus may not be ready today; however, the commercial sector is working to solve many of the issues of importance to training such as resolution, field of view, and tracking latency. Research is needed to determine the qualities required for particular uses of low-cost, commercial virtual-reality technologies in training. It is recommended this research be performed in parallel with commercial technology development.

Commercial gaming technologies provide an engaging entertainment environment, rewarding the player for demonstrated competencies—many of which are learned within the game. To benefit from learning afforded by approaches used in gaming, research is needed to identify training tasks most appropriate to utilize game technologies to impart transferrable skills.

Performance Assessment

A wealth of performance data can be captured—physiology data, trainee input data, system performance, outcomes, etc. This data can be analyzed against various performance metrics and utilized as an instructor aid or for instructorless training across multiple ranges of device fidelities to provide feedback on performance or adapt the learning to the student.

Integrated—Live Virtual Constructive

The vision for I-LVC includes the entire kill chain, including C2 and national assets. Both live red and blue assets can be supplemented with virtual and/or constructive participants. All participants appropriately sense and communicate with other participants seamlessly across the L-VC boundary, with the ability to launch air and ground constructive weapons with real-time scoring and kill removal. Instructors have the ability to assume the role of constructive threats to be able to introduce the human element when required. A constructive environment server provides a robust environment, and ground-based tools provide the common operating picture and debriefing capability.

Boeing's foundational integrated LVC research began in 2007 with a live F-15E, a virtual F-15E, and constructive red air in a blue verses red engagement. Progressive development and demonstrations added multiple capabilities for both air-to-air and air-to-ground on the F-15E and expanded to the F/A-18E/F. As a result, industry is ready to deliver I-LVC solutions today. It is recommended the Air Force aggressively pursue an acquisition program to realize demonstrated benefits. Research is needed to determine modifications to live training to realize the maximum benefit from I-LVC, along with targeted developments of credible constructive opposing force and sensor models for certain training tasks.

AFRL/RHAS believe that distributed LVC training can bring down those training costs and improve training effectiveness just as LVC distributed mission operations training has brought down training costs and improved training effectiveness in the aviation community.

BOX 2-4

CAE, Inc.

David Graham, Senior Technical Fellow

CAE, Inc., is a publicly held, independent, medium-sized company with products and services focused on the creation of domain expertise using modeling and simulation. Our business is roughly half military and half "civil," and our business is also roughly half supplying products and half providing services. CAE is honored to have the opportunity to provide our industry perspective on promising new approaches to employment of simulation in the U.S. Air Force Training Environment.

The CAE presenter will briefly review CAE's current products and services in use by the Air Force and other end users and respond to the questions about what we are doing now and what the shortfalls of current simulation industry offerings and technology are.

CAE's view of what we would like to be doing and what it will take to achieve our ambitions will be collected in two broad categories: "not-so-thin" simulation clients and "thin" simulation clients.

"Not-so-thin" is one way to describe high-performance, full-flight simulators that make up a very large part of CAE's product and service offerings to both civil and military customers. Promising new approaches will focus primarily on increasing the capability to interoperate federations of heterogeneous simulators to improve the capability to use simulators for mission sets that AMC accurately describes as "not optimal" in their presentation abstract. The CAE presenter will explore the role of open, consensus-based standards to help achieve the promise that rapidly advancing technology can potentially deliver.

CAE believes there is a very promising future in the use of simulation viewed through "thin" clients: zero-deployment web-browsers on a wide variety of hardware and software platforms. New learning sequences that expose training audiences to simulation at various levels of detail and complexity are becoming possible and offer the promise of low-cost, low-risk, rapid expansion and connectivity of elements of mission management components to distributed mission training and rehearsal events. In addition, the capability to "bring the high-performance simulation software to the desktop or mobile device" offers the promise of new, dynamic, highly engaging learning sequences in what we have traditionally considered "ground school."

The presentation will conclude with a discussion of collaboration between U.S. Special Operations Command and the Joint Staff / J7 in the JLVC 2020. A brief examination and demonstration of the J7 Cloud Based Terrain Generation Service will serve to integrate the points previously discussed and support specific recommendations by the CAE presenter.

BOX 2-5
FlightSafety International

Nidal Sammur, Director of Engineering

FlightSafety International has long believed the best safety device in any aircraft is a well-trained crew. To that end, we have continually invested in technology and training innovations that provide the highest possible fidelity training to our customers, both commercial and military. In support of that objective, FlightSafety is focused on designing, manufacturing, and sustaining high-fidelity training devices intended to offer the most realistic immersive training environment possible. Our presentation will address the current state of technology in simulation, explain initiatives we are currently pursuing, and posit future areas for innovation, all with an eye towards continuing to enhance the realism of the training experience of our customers.

BOX 2-6

Federal Aviation Administration

Jeffery A. Schroeder, Chief Scientific and Technical Advisor
Flight Simulation Systems

The Federal Aviation Administration (FAA) regulates simulators for pilot training and uses simulators to train air traffic controllers, site new control towers, design airspace procedures, and develop unmanned aircraft systems requirements. This presentation focuses on simulators for airline pilot training only. Piloted simulation represents the largest and most sophisticated component of the FAA's responsibility in simulation, and these simulators must comply with federal regulations before they are used in pilot training. Airline pilots fly the simulator once or twice per year for about three days. Most of that time covers mandated training items, but an airline typically adds specialized training deemed important based on analysis of their operations. Once a year, pilots must pass a proficiency check in the simulator. The accident rate in the United States suggests that this process is satisfactory, as the rate continues to decrease with the continued increase of simulation use.

Naturally, these simulators still have limitations. This limitations fall into two categories: (1) the device is not capable, or (2) the device is capable, but is not used for the purpose. The latter category is not a limitation of the device itself, but of its application. Instances in the first category include (a) fully simulating the environment outside of the aircraft such as air traffic control and surface vehicles; (b) the lack of in-flight surprise; (c) motion cueing differences, especially normal and lateral load factors; (d) poor fidelity in wake vortex encounters; (e) stall modeling; (f) physical effects of icing; (g) stability and control fidelity near envelope edges; and (h) the landing experience is still different from flight. Items in the second category include (i) not demonstrating some key pilot-vehicle interface functions and (ii) simulating events in conditions that differ from those that typically occur in flight (e.g., go-arounds, stalls).

Besides trying to improve the above limitations, additional simulator enhancements may further improve aviation safety. These enhancements include (1) being able to get yesterday's incident into training instantly to prognostically prevent tomorrow's accidents; (2) developing scenarios that invoke grey decision making and that expose common human errors; (3) defining the relation between simulator fidelity and training value; (4) adjusting the challenges posed in simulation to be commensurate with the trainee's skills; (5) relying more on frequency-domain measures to ensure that the simulator and aircraft have similar flying qualities; and (6) better modeling of slippery runway conditions.

As far as technologies, approaches, and techniques required to satisfy this to-do list go, much of it is simply time, money, and the will power to do it. Many of the improvements are evolutionary instead of revolutionary. Probably a lot can be done with standardization so that improvements can be made more collectively, rather than in an individual piecemeal approach. However, incentives to standardize and the enthusiasm for doing so have not been self-evident. Also, the pressure to keep training costs manageable necessitates that hard decisions be made on what not to do if more is added to a training session.

BOX 2-7

National Aeronautics and Space Administration

Bimal Aponso, NASA Ames Research Center

NASA Ames Research Center is home to several high-fidelity research flight and air-traffic control simulation facilities which, together with an experienced workforce, produce high-quality research data and findings that have proven to be applicable in the real world. These assets include the Vertical Motion Simulator (VMS), Crew Vehicle Systems Research Facility (CVSRF), Future Flight Central (FFC) air traffic control tower simulator, and several air-traffic control (ATC) simulators.

The VMS combines a high-fidelity simulation capability with an adaptable simulation environment, enabling customization for numerous human-in-the-loop research applications. The distinctive feature of the VMS is its unparalleled large-amplitude, high-fidelity motion capability. In over 30 years of continuous operation, the VMS has contributed significantly to the body of knowledge in a range of disciplines directly benefiting several aerospace programs and flight safety, including the design and development of flight control systems for the Joint Strike Fighter, Space Shuttle Orbiter, and rotorcraft. It continues to be used for researching new vehicle configurations, vehicle control and safety, transfer-of-training, etc., by NASA, other government agencies, and industry.

The CVSRF includes two motion-based flight simulators: a Boeing 747-400 full-flight simulator and the reconfigurable Advanced Concepts Flight Simulator (ACFS). These simulators are primarily used to research air-traffic management concepts and procedures, advanced navigation and avionics concepts, and cockpit human factors. FFC is a full-sized control tower simulator with a 360-degree external field-of-view display system and reconfigurable system architecture. FFC and the ATC simulators are used to test air-traffic management automation and decision support tools and demonstrate their feasibility in a realistic environment prior to technology transfer for implementation in the National Airspace System.

To support integrated simulations and flight tests for NASA's Unmanned Aircraft Systems (UAS) in the National Airspace System Project, NASA developed a distributed test environment incorporating live, virtual, constructive (LVC) concepts. Development of the software enabling the LVC is conducted primarily at the Distributed Simulation Research Lab at NASA Ames. The LVC components provide the core infrastructure supporting simulation of UAS operations by integrating live and virtual aircraft in a realistic air-traffic environment. This provides the ability to conduct tests more efficiently by promoting the use of existing distributed assets. The LVC infrastructure was used in several human-in-the-loop simulations to evaluate acceptance of Detect and Avoid advisories used by UAS pilots to maintain well clear of other virtual traffic and to negotiate maneuvers with air-traffic control. It is currently being used to support testing of self-separation algorithms between unmanned and manned aircraft in live flight. Further simulations with more comprehensive air traffic scenarios mixing live and virtual aircraft is planned.

In the current fiscal environment, maintaining and upgrading these high-fidelity simulation assets and retaining the skilled workforce necessary to meet future research needs is the primary non-technical challenge. Technical challenges include the ability to develop and participate in LVC-distributed simulations more quickly and with less cost expenditure on developing customized solutions. Potential solutions include determining and establishing interface definition standards for interacting simulation environments covering simulation models, communication protocols, information technology security, etc. Also, an improved understanding of the benefits of simulation and levels of simulation fidelity required for program risk mitigation and training effectiveness would better inform funding decisions on these assets.

BOX 2-8

The University of Toledo Interprofessional Immersive Simulation Center

Pamela Boyers, Executive Director, University of Toledo Interprofessional Immersive Simulation Center; Gerald Zelenock, Professor and Chairman, Department of Surgery, University of Toledo College of Medicine

The University of Toledo Interprofessional Immersive Simulation Center (UT-IISC) is a highly advanced 65,000 sq. ft. simulation facility purpose-designed to transform the training of health care providers and develop new methods for improving human performance and effectiveness. With a unique clustering of three highly integrated, state-of-the-art simulation centers, the UT-IISC provides the ideal venue in which medical/industry partnerships are created for the purpose of developing and testing of new processes, products, and devices. In addition, UT-IISC has a wide range of subject matter experts available to advise, support, and help test the development of new products—including the potential of partnering to conduct human factors research and develop autonomous health systems.

A Tri-Center Simulation Training Concept

The UT-IISC houses three distinct, yet integrated, simulation centers:

- A *Modeling and Simulation Center* that incorporates 3-D and Virtual Immersive Reality (VIR) and holographic technology with a 5-sided light-emitting diode (LED) VIR, a large, curved LED CAD Wall, a Holographic Theater, Display Wall, and Industry Collaboration Spaces.
- An *Advanced Simulation Center* that houses real hospital equipment and human patient simulators in a wide variety of simulated healthcare settings—including an *Elliptical Virtual Hospital* that incorporates an Intensive Care Unit, Labor and Delivery Room, Trauma Suite, and a Pediatric Unit around a central control tower. The human patient simulators are computer "driven" through medical scenarios from this control room that is surrounded by one-way glass. This design enables the simulation scenarios to be easily viewed from a raised vantage point. All virtual clinical environments have cameras and microphones installed in the ceilings to record each training session. Critical events that occur during the LVCEs are tagged by the simulation capture system and participants review the exercise in adjacent debriefing rooms utilizing audio and visual recordings—along with the physiological data (clinical responses) of the human patient simulators.
- An *Advanced Surgical Skills Center* containing 17 surgical bays and procedural rooms is equipped with advanced surgical equipment that includes up-to-date instrumentation and a wide range of surgical scopes. The center operates in partnership with surgical instrumentation companies who help support the learning and research activities by providing equipment and staff for procedural skills and product development workshops.

From both the training and research and product development perspectives, it is possible to use all three centers to achieve the desired objectives. For example, one can "fly through" a human heart using the VIR in the Modeling and Simulation Center, then practice conducting a "Code Blue" as a team member in the Advanced Simulation Center, followed by conducting cardiac procedures in the simulated surgical suites in the Advanced Surgical Skills Center.

Promoting interdisciplinary collaboration and human factors research, the UT-IISC supports the development of procedural and communication skills through the ongoing development of reliable, valid methods of competency assessment. The ultimate goal for the UT-IISC is to focus on how simulation and LVC exercises in replicated clinical settings can improve the outcomes of care through enhancing the efficiency and accuracy of individuals and teams—ultimately reducing the costs of healthcare.

To transform the education of health professionals, the UT-IISC is utilizing a convergence of advanced simulation technology to help break down barriers (stove pipes/silos) between professions by promoting collaborative practice and using simulated clinical scenarios to enhance the performance of individuals and teams. The overarching mission of the UT-IISC is improving healthcare outcomes—with a strong emphasis on improving patient safety. The wide spectrum of modeling and simulation modalities available in the UT-IISC place the University of Toledo in a position to utilize "disruptive technologies" to transform the medical learning and research environment. Through the provision of interdisciplinary simulation and clinical simulation experts, the UT-IISC welcomes collaboration with many disciplines, including the U.S. military, in improving the outcomes of training and the design and testing of new products, processes, procedures, and systems.

BOX 2-9

State University of New York at Binghamton

Frank Cardullo, Professor of Mechanical Engineering

The presentation aims to illuminate some of the flight simulation technology areas that present potential obstacles to successful pilot or other crewmember training. The simulator is discussed as a complex, dynamic, man-machine system in which the human operator is central to achieving the goals of exercise. It will treat technology issues of dynamic system simulation, human perception, and behavior in the context of a control theoretic approach. A major advantage of this approach is that, if applied appropriately, it will yield quantitative metrics of the simulator as a training device. It has been demonstrated that when certain anomalies occur in a flight simulator, such as visual or motion artifacts or the absence of certain cues necessary for proper execution of the task, that pilot performance metrics may remain constant but control behavior is altered. The discussion will include an introduction to some of the signal-processing techniques that can be used to quantitatively analyze pilot control behavior. Some examples will be presented, such as in the case of uncompensated delay in the various dynamic systems and the Objective Motion Cueing Test recently developed that quantifies in the frequency domain the effects of the motion cueing algorithm on the total motion system dynamics. The talk will conclude with some suggested areas of development.

3

Committee Member Observations on Adapting Additional Simulation Techniques for the Air Force

AREAS WHERE THE AIR FORCE COULD BENEFIT FROM ALTERNATIVE USES AND TECHNOLOGIES

A plethora of observations resulted from the presentations in Chapter 2. This section begins with extensive observations regarding future simulation architectures (Box 3-1, Box 3-2, and Figure 3-1, with associated explanation).

Committee member observations touched on the broad concepts above. First, Don Fraser, committee co-chair, and several other committee members were optimistic that, based on the earlier presentations, a significant part of this architecture concept is already in place (e.g., in the distributed mission operations network known as DMON). These committee members noted that movement forward can thus evolve in stepwise fashion with advances sized to meet specific training needs. (Note: Col Nathan Hill, Chief of ACC Flight Operations, mentioned issues in this area: "How many networks are too many? What type of networks do we need? What are the second and third order effects of shutting down and consolidating networks?") John-Paul Clarke opined that the Air Force needs a modular-flexible framework as a strategy on which to hang tactics and mechanisms to promote convergence versus a large program of record. He went on to say that it is necessary to know what standards to use. John Corley offered that the development of an intellectual architecture for live, virtual, constructive (LVC) simulation must occur prior to contracting for the physical architecture. Mr. Corley supported the use of the Drew-Robinson architecture concept. Mr. Corley believes the intellectual construct should not demand investment but provide a framework for decision makers to "opt in" where LVC supports learning opportunities not available through other methods, or where value is enhanced. He stated that the system design must be sufficiently adaptive to delivery of knowledge that, on the whole, delivers learning that is more rapidly assimilated and retained for longer periods.[1]

Committee co-chair Ray Johns offered that having established standards will allow the Air Force to have lower life-cycle costs. In a related topic, Michael Zyda noted that the U.S. government has failed miserably in simulator standards. He said, "Why not use open source procedures and processes?" Ex-post facto standards are hard to do, and very expensive. Dr. Zyda stated that the National Research Council's 1997 report *Modeling and Simulation: Linking Entertainment and Defense*, which he chaired, raised almost all the same issues with respect to the internetworking of defense simulations.[2] The lengthy architecture dialog led to the following additional key themes.

---

[1] Bimal Aponso, Chief, Aerospace Simulation Research and Development Branch, a suggested using a phased approach to developing the common architecture using limited operational scenarios. The stated aim of this approach is to reduce the risk of integrating LVC components. Large-scale demonstrations and tests are inherently difficult to assess in terms of effectiveness due to the sheer scale of the variables involved. A phased build up to a large scale test using smaller, easier to measure, operationally relevant scenarios may be a better approach.

[2] For additional information, see National Research Council. *Modeling and Simulation: Linking Entertainment and Defense*, Washington, D.C.: The National Academies Press, 1997.

BOX 3-1

Observations on Path Forward for Integrating Air Force LVC Efforts

Pamela Drew, Committee Member

1. The implementations of live, virtual, constructive (LVC) simulation for training currently underway in the Air Force, Navy, and elsewhere are being developed in independent, stovepipe, and ad hoc fashion, which results in a platform-centric capability with simulator-simulator (hardwired) interfaces, disintegrated networks, and duplicative and similar, but unstandardized and unshared, data and mission sets. An alternative, and what is needed, is an approach that creates a common architectural approach in which LVC simulations can be "plugged" into an integration LVC backbone or integration architecture—hereafter referred to as ILVC-IA. Figure 3-1, from co-member Harry Robinson, illustrates this type of architecture.

a. In this architecture, there would reside reusable data for terrain, weather, threat information, blue tasking, etc. It would also contain reusable mission models, mission logic and rules, and simulators that could be re-purposed and used in various applications or instances of ILVC training sessions. The live or VC simulations would be integrated into this environment via standardized interfaces for communications and data links, for SIM via DIS and HLA, and the data passed would have to conform to standardized access interface protocols. Using this common integration architecture and enforcement of standards, a proprietary solution can still be integrated as long as it conforms to the interface and data access requirements.

b. This architecture can be put into use to support the entire range of desired combinations of LVC to support all missions from the "high end" Combat Air Force (CAF) requirements to more routine VC training scenarios. These mission scenarios create use cases of the architecture and results in specific applications or instances (e.g., an F-35 Live pulling VC world view of KC-46, AWACS, weather, etc.).

c. Of note, real-world sensors can also be integrated as feeds into the system, thereby bringing "reality" to the simulation. Obvious examples are for terrain and weather as part of the "live" feed, as well as other live assets.

d. Finally, security was referenced in multiple ways as a gap or obstacle by various presenters. In the ILVC-IA, security would have to be addressed. A few different elements would include encryption for the transport layer; multilevel security for crossing classification levels; role-based, access-control-type capability for authentication and authorization; and physical security for facilities.

2. By creating this new architecture, it would be possible to transform from a platform-centric view to a reality-centric view, enable more rapid integration of simulated and live assets, and enable far more efficient development of training capabilities.

Theme 3. Currently, live, virtual, and constructive training efforts are evolving in a largely ad hoc, stovepiped, and somewhat inefficient fashion. This situation suggests Air Force consideration of a different architectural approach that would be world-centric—open, pluggable and playable—rather than platform and contractor proprietary centric. This world-centric construct would contain common elements and live data, such as weather, terrain, threats, with an array of specific simulation platforms around the periphery drawing information from the common databases as opposed to utilizing their own proprietary database (Pamela Drew, Harry Robinson).

Theme 4. There are indications that some elements of the Air Force simulation architecture currently have these world-centric enterprise characteristics, so continued pursuit of an enterprise-level solution to live, virtual, constructive training could be very beneficial (Pamela Drew, Harry Robinson).

3. While this can be viewed as a technical architecture, the Air Force sponsors see it as providing a framework to articulate potential investment needs and to prioritize "where the next dollar should be spent." CAF and Mobility Air Forces (MAF) representatives both commented that, of the data sets presented during our general discussion, geographic, terrain, and threat sets were the priority.

4. CAF has an emergent and urgent need to bring VC simulation to augment F-35 Live to enable training due to constraints stated in the workshop. These are a combination of the decision not to allow full capability in live training, amongst others.

5. There is a need to organize the development of such an architecture through a clear authority structure, which would lead the architecture, standards, interface, and reusable asset-data-capability development. Note the goal should be to leverage all that can be reused or adapted to that which already exists.

6. The advanced technology demonstration (ATD) presented by Wink Bennett (Air Force Research Laboratory, AFRL) is an example of one "bottoms-up" instance of LVC underway. This could be harnessed and adapted as needed to become a first instance to begin implementation of the ILVC-IA architecture.

7. Just as such an architecture would benefit the Air Force, there is an analogous gap and application across the services—Department of Defense (DoD) wide. The Navy is also just beginning the LVC journey, developing yet another (mostly separate) capability operating on the JBUS, which appears to be the counterpart to the Air Force distributed mission operations network (DMON). Getting the services to use the Defense Information Systems Agency Global Information Grid (DISA GIG) via the Joint Information Environment (JIE) will facilitate the transport/network layer of integration.

8. There are various efforts underway that address some or perhaps all of the proposed ILVC-IA. These include the J7's JLVC Vision 2020, the AFRL ATD, Air Force Special Operations Command Ops training, and industry efforts (e.g., Boeing, Lockheed, Northrop capability). These efforts should be assessed and leveraged into this unified ILVC-IA capability as possible and appropriate.

9. There is a need for a single authority within the Air Force to define architecture, enforce standards, to select and maintain reusable content of the ILVC-IA, including, but not limited to, reusable data and mission sets. The authority should also create and drive execution against a near-, mid-, and long-term roadmap and associated plan that demonstrates capacity to integrate legacy capabilities (both government and industry) with new capabilities. In addition, and as important, are a new governance model, communication model, and stakeholder engagement.

10. There were a variety of technology developments and improvements for human-in- the-loop interfaces (e.g., Google Glass) and techniques (e.g., motion) that can be included in a continuous technology refresh sub-task in the oversight and development of the ILVC-IA. These assessments must also be specific to training objectives.

Committee members had additional observations in other areas. First, Robert Allardice noted that, in connection with the Boeing presentation, a benefit is that current and emerging technology for assessment and gaming technology may provide significant growth in our understanding of learning. He also noted that mobile technology has changed how people make decisions; we ought to heavily leverage mobile technology for enhancing learning and substituting training. Mr. Allardice went on to say that more discussion should take place on what the Air Force understands about "how" humans can best learn "today" based on significant discoveries and advances within the past decade. He said it is important to tailor the right learning tool for the right learning objective and place competency in the right platform. Finally, Mr. Allardice shared that, regarding the medical presentations, there are tremendous lessons to learn from collaborating with the medical community. That community, he noted, is advancing understanding of learning, education, and training, making significant advances in several technologies that could help the Air Force prioritize and match content to learning platforms. Leveraging technology to deliver an experiential learning environment similar to medical simulation is important.

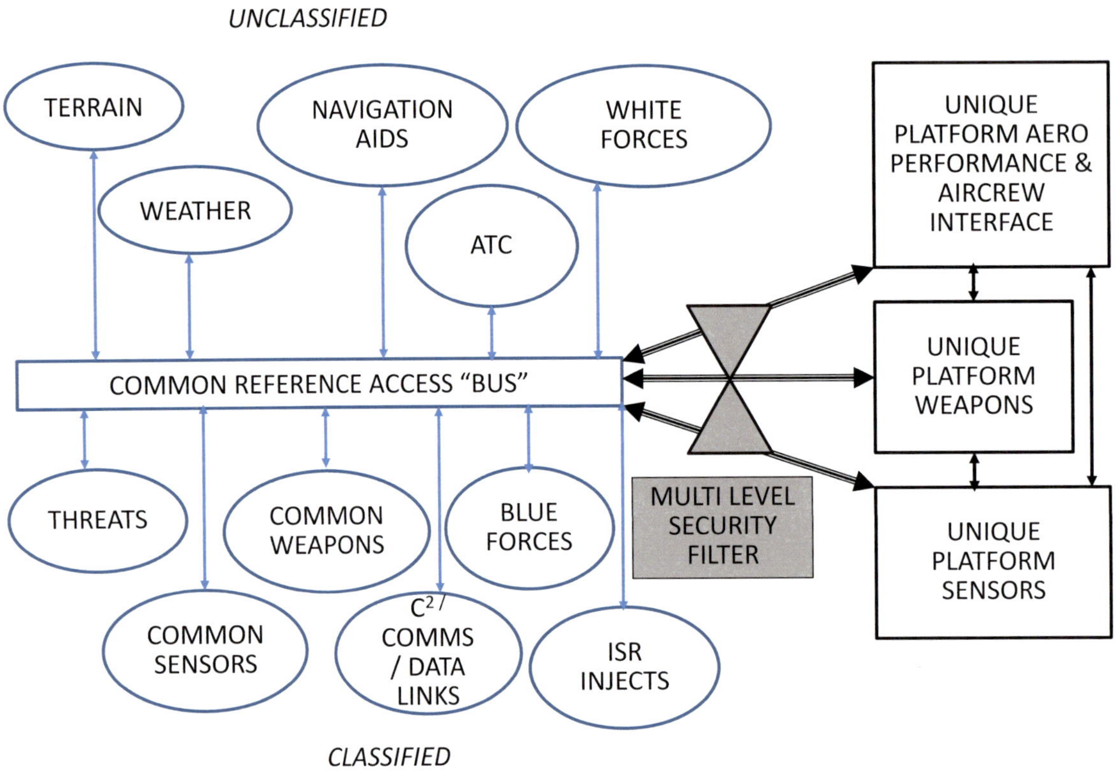

FIGURE 3-1 Notional architecture for U.S. Air Force live, virtual, constructive training.

BOX 3-2

Explanation of Notional Architecture for U.S. Air Force LVC Training

Harry Robinson, National Program Manager, Veterans Health Administration (VHA) Simulation Learning Education and Research Network (SimLEARN) (Committee Member)

Simulation-based training environments for the Air Force would benefit from an architecture using Common Reference Access "Bus" that would serve as a shared information provider simultaneously supporting generation of mission characteristics and events necessary to provide realistic training. Components would contain grouped databases that would drive LVC simulations accessed from training platforms unique to specific aircraft types, models, and series. Each component database would be established and subsequently maintained to achieve necessary level of currency. The respective modules could be characterized as Unclassified (including terrain, weather, navigational aids, air traffic control, and white force generation) or Classified (including threat, enemy orders of battle, common sensors [similar across multiple aircraft], common weapons [air-to-air, air-to-ground, similar across multiple aircraft], command and control, communications, data links, blue force generator, and intelligence-surveillance-reconnaissance scenario injects. Simulation execution of unique platform models for aerodynamic performance, aircrew interface (e.g., controls and displays), weapons, and sensors would integrate with the components' data accessible on the common reference access "bus" as controlled or limited by a multilevel security filter. Specific aircraft simulators would plug in to the common bus. Advantages of this construct include (1) reduction in need for stove-piped and proprietary solutions for each type aircraft simulator, (2) standardized component databases that can be independently established, (3) ease for maintaining database currency, (4) networked simulations executed in a shared environment, and (5) simulation-based exercises that support specific platform security program requirements.

John Corley offered that the ability to deliver learning for training is important and that consideration should be given to the changes in how airmen will "learn" and the related future demographics. (Note: Col Hill also had a comment in this area: "The new generation learns differently than most of us. We need to figure out the best way to teach them.") Steven Detro observed that, as a substitute for some training, higher-fidelity simulation technologies are now enabling more training to be accomplished in virtual reality. He went on to note that continued analysis of the potential benefits that virtual reality simulation could offer to each area of training should be considered. With a blend of different training media and training devices, Mr. Detro offered that a greater percentage of training sorties or training events could move into simulators; these hours should complement current live fly hours. (Note: Current acquisition policies have forced the Air Force to find lower-cost technology.) Mr. Detro believes the Air Force could use (1) performance measurement technologies already developed by the Air Force Research Laboratory (AFRL) to increase the ability to objectively measure the effectiveness of training;[3] (2) the "science of learning" cognitive modeling products of the AFRL to assist in the development of more efficient learning delivery methods for training; and (3) immersive technology advancements to deliver training information to match the learning preferences of students.

Several committee members provided final thoughts in this area. First, Ray Johns offered that independent research and development by industry has advanced knowledge of simulation applications and technologies. Second, Harry Robinson noted that, to meet Air Force needs, the following outside capabilities and technologies are most useful: adaptive learning and intelligent tutoring, cloud computing, common accessed resources for data, real-time representations and feedback loops that avoid latency issues, and multiplayer interactive gaming that builds teamwork and communication skills. Lastly, Michael Zyda noted that, with respect to the CAE presentation, open standards for all parts of the simulation enterprise will decrease costs and make better systems. He said that, regarding the Lockheed Martin presentation, alternate simulation systems become possible with head-mounted displays, and perhaps the Air Force should look at head-mounted displays and augmented reality technology for some of what it is doing. He also thinks that networked simulators have latency problems; perhaps look at what the game industry does for this.[4]

HOW LIVE, VIRTUAL, CONSTRUCTIVE TECHNIQUES COULD IMPROVE AIRCREW TRAINING

Several committee member comments applied to how techniques for LVC simulation for training could improve aircrew training, although there are links back to other messages in this report. For example, Robert Allardice noted that at one point simulator training was secondary; however, advances in technology have led to LVC as the primary way to train for the mission. Mr. Allardice went on to say that he thinks the best way to frame LVC is not that it will improve training. It is that advances in technology and modern applications drive an LVC "imperative." Mr. Allardice noted that all training can benefit to some extent; the key seems to be to develop an architecture from which specific applications can draw, based on the risk profile the Air Force chooses based on a particular mission set. John Corley was of the opinion that the Air Force needs to make prudent investments that enable needed enhancements to or development of the enterprise intended to yield a "realistic" training environment. Mr. Corley noted that

---

[3] There was a comment from an operational pilot in the audience on the need to measure training effectiveness. The pilot said that when developing a training capability, particularly of threat environments, it was important to ensure that the probability of success in the simulator be equivalent to that in an actual situation. This highlights the overall issue of ensuring a viable training effectiveness validation method is developed in tandem with LVC simulation capability.

[4] For best practices, consider the following publication: Johns Hopkins University Applied Physics Laboratory, *Best Practices for the Development of Models and Simulations: Final Report*, NSAD-R-2010-037, Laurel, Md., June 2010, available at http://www.msco.mil/MSBPD.html.

investments must consider the temporal dimension, bit-sized approach toward the delivery of the LVC capability. The following themes arose during the discussions:

> **Theme 5.** Advances in technology and increasingly complex user needs have led to live, virtual, constructive training as the primary way to train for some missions (Robert Allardice).

> **Theme 6.** Substantial benefits could accrue to the Air Force if it relied on open systems and acquired data rights as the model when procuring new systems. Enforcing compliance to more interoperable, related standards could lead to a "plug and play" environment (Pamela Drew, Michael Zyda).

> **Theme 7.** Research into the "science of learning" is indicating that young people, who have considerable computer skills compared to previous generations, learn in very different ways compared to older generations. Future architectures and systems would benefit by taking this knowledge into account (adaptive learning) (Donald Fraser, Steve Detro).

SUGGESTED AREAS FOR POSSIBLE FOLLOW-ON STUDY

Ray Johns indicated that the Air Force sponsors of this workshop requested that there be no follow-on study. Nevertheless, some committee members suggested a few areas that the Air Force may wish to delve into more deeply; these areas are listed below.

1. What is the full set of requirements for Air Force LVC simulation for training?
2. What is the optimal standard and architecture that the Air Force should strive for? What is the roadmap for the architecture?
3. How can multilevel security be dealt with—through a study in its own right? and Should such a study be classified?
4. What can be done about adaptive learning?
5. What is the need, if any, for a change in Air Force governance with respect to LVC simulation for training? What organizational and budget changes need to be made for an effective LVC simulation for training capability across all missions (with the F-35 as the first system priority)?

Appendixes

A

Biographical Sketches of Committee Members

RAYMOND E. JOHNS JR., *Co-Chair,* is responsible for FlightSafety International's global government and military programs. He began working with the company in 2013 as a senior advisor and was named senior vice president in January 2014. Before assuming his current role, Gen. Johns commanded the United States Air Force Air Mobility Command, Scott Air Force Base (AFB), Illinois. The mission of the Air Mobility Command is to provide rapid, global mobility, and sustainment for the U.S. armed forces. Gen. Johns graduated from the U.S. Air Force Academy in 1977. He has served as a program manager and source selection authority; an experimental test pilot, having flown some 83 different aircraft; and he was the chief test pilot and test program manager for the VC-25 Air Force One Replacement Program. He was chosen as a White House fellow in 1991, where he was a senior staff member in the Office of National Service. He served at Headquarters US European Command in security assistance, strategy, and congressional affairs and at Headquarters US Pacific Command as deputy director of strategic plans and policy. He commanded a test squadron, operations group, and airlift wing, and he was the director of mobility forces for operations in Bosnia and was responsible for strategic airlift operations in Iraq and Afghanistan. Gen. Johns served as deputy chief of staff for strategic plans and programs, Headquarters U.S. Air Force, Washington, D.C., where he developed, integrated, evaluated, and analyzed the U.S. Air Force annual budget and the Air Force Long-Range Plan to support national security objectives and military strategy. He retired from the U.S. Air Force effective January 1, 2013.

DONALD C. FRASER, *Co-Chair,* has broad research and development management experience and is the founder and retired director of the Boston University Photonics Center. Dr. Fraser has had a distinguished career managing the development of high technology enterprises, both in the private and public sectors. Dr. Fraser joined the Massachusetts Institute of Technology's (MIT's) Instrumentation Laboratory (which became the Charles Stark Draper Laboratory in 1973) as a member of the technical staff; later he served as the director of the Control and Flight Dynamics Division; vice president of technical operations; and executive vice president and chief operating officer. From 1990 to 1991, Dr. Fraser was deputy director of operational testing and evaluation for command, control, communications, and intelligence at the Department of Defense (DoD). After Senate confirmation, he was appointed Principal Deputy Under Secretary of Defense (Acquisition) from 1991 to 1993. From 1994 to 2006, Dr. Fraser was the director of the Boston University Photonics Center and a professor of engineering and physics. His honors include the Defense Distinguished Service Medal. Dr. Fraser has served on the NASA Advisory Council. He is a member of the National Academy of Engineering, served on the National Research Council's (NRC's) Aeronautics and Space Engineering Board, chaired several NRC committees, and was a member of many other NRC committees. He received his Sc.D. in instrumentation from MIT.

ROBERT R. ALLARDICE founded Allardice™ Enterprises, Inc., in 2013 after successfully serving in the U.S. Air Force for more than 33 years, reaching the rank of Lieutenant General. With 16 years of senior executive experience and a remarkable record of achievement in the Air Force, Mr. Allardice is recognized as an innovative pioneer leading transformation in modern complex global systems. His experience leading organizations ranging from 100 to 133,000 people, culminated in the position of vice commander of Air Mobility Command (AMC). In that capacity, he ran corporate oversight of a $20 billion operation with broad responsibilities from operations and training to programming, installation oversight, and financial management. Additionally, he sat on the U.S. Air Force Corporate Board and several operational governance boards. Prior to his duties at AMC, as commander of 18th Air Force, he led the U.S. military global air mobility enterprise through transformation and multiple global operations. His leadership of the military's global air transportation system is credited with unique applications of virtual and collaborative tools redefining modern staffing methods and driving significant increases in effectiveness and efficiency. Prior experiences include oversight of the U.S. Central Command Security Assistance program for Central Asia, and the Mideast, working with 20 different countries to refine security cooperation agreements. Also, he led the team building the U.S. military strategy for the Mideast, Central Asia, and Persian Gulf. Additional recent experience includes command of the Coalition Air Force Transition Team in Iraq, where he successfully built a program to reestablish the Iraq Air Force. Mr. Allardice holds an M.S. in systems management from the University of Southern California.

JOHN-PAUL CLARKE is an associate professor in the Daniel Guggenheim School of Aerospace Engineering, with a courtesy appointment in the H. Milton Stewart School of Industrial and Systems Engineering, and director of the Air Transportation Laboratory at the Georgia Institute of Technology. He received S.B. (1991), S.M. (1992), and Sc. (1997) degrees in aeronautics and astronautics from MIT. His research and teaching in the areas of control, optimization, and system analysis, architecture, and design are motivated by his desire to simultaneously maximize the efficiency and minimize the societal costs (especially on the environment) of the global air transportation system. Dr. Clarke has made seminal contributions in the areas of air traffic management, aircraft operations, and airline operations—the three key elements of the air transportation system—and has been recognized globally for developing, among other things, key analytical foundations for the Continuous Descent Arrival and novel concepts for robust airline scheduling. His research has resulted in significant changes in engineering methods, processes, and products—most notably the development of new arrival procedures for four major U.S. airports and one European airport—and changes in airline scheduling practices. He is an associate fellow of the American Institute of Aeronautics and Astronautics (AIAA) and a member of the Airline Group of the International Federation of Operations Research Societies, the Institute for Operations Research and the Management Sciences, and Sigma Xi. His many honors include the AIAA/AAAE/ACC Jay Hollingsworth Speas Airport Award in 1999, the Federal Aviation Administration (FAA) Excellence in Aviation Award in 2003, the National Academy of Engineering Gilbreth Lectureship in 2006, and the 37th SAE/AIAA William Littlewood Memorial Lecture Award (awarded in January 2012).

JOHN D.W. CORLEY is an experienced strategic thinker and skilled international collaborator in the development and utilization of weapons systems. He entered the U.S. Air Force Academy in 1973. His aviation career includes more than 3,000 flying hours with combat experience. He commanded at flight, squadron, group, wing, and major command levels. His staff positions comprised a mix of service and joint duties in Tactical Air Command, Pacific Air Forces, U.S. Air Forces Europe, Air Combat Command, Headquarters U.S. Air Force, and the Joint Staff. Gen. Corley retired from the U.S. Air Force after 36 years on active duty. His final assignment was commander, Air Combat Command (ACC). At ACC, he directed the planning, organizing and training to assure combat-ready forces for 156,000 personnel operating 1,200 aircraft at more than 200 worldwide locations. He orchestrated the development of strategy, doctrine, concepts, and procedures for air power employment. Previously, he served as vice chief of staff, U.S. Air Force, responsible for the oversight of 680,000 active-duty, Guard, Reserve, and civilian personnel serving in the United States and overseas. Other key staff positions included the following: principal deputy, assistant secretary of the Air Force for acquisition; military director, member of the U.S. Air Force Scientific Advisory Board; and director, studies and analysis, Headquarters U.S. Air Forces in Europe. Since retiring from active duty, Gen. Corley has become an independent consultant. He serves on several boards in addition to consulting for a number of defense and aerospace industry corporations. He served on the board of the Air Force Association to educate the public about the critical role of aerospace power in the defense of our nation, advocate for aerospace power, and support the Air Force family and aerospace education. Additionally, he is a trustee of the Falcon Foundation, providing scholarship funding for promising young men and women aspiring to attend the U.S. Air Force Academy.

STEPHEN D. DETRO directs a team at Lockheed Martin in new business forecasting, business capture, and marketing activities focused on domestic and international simulation and training opportunities. Mr. Detro has more than 35 years as a business development executive representing companies and leading multi-disciplined teams providing simulation and training technologies and solutions for DoD and international Air Forces. He is a retired lieutenant colonel from the U.S. Air Force Reserve and Air National Guard, with 28 years total service with sustained combat mission ready status as a U.S. Air Force Reserve and Air National Guard fighter pilot, while maintaining a full-time civilian career. He is a combat mission-qualified pilot in the F-16A, F-4D, A-7D and F-100D fighter aircraft and a command fighter pilot with more than 2,300 hours and 4 years of enlisted service in aircraft maintenance. Mr. Detro is also currently protocol officer and conference chair emeritus for the Interservice/Industry, Training, Simulation and Education Conference and former chairman of the National Training and Simulation Association Executive Committee. Mr. Detro holds a B.S. in education from Wright State University.

PAMELA (PAM) DREW is executive vice president and president of information systems, a business area of Exelis, Inc., that is a leading provider of mission critical network solutions. These solutions leverage the group's core capabilities that span the full life cycle of critical networks, including system architecture, design, development, deployment, integration, test and evaluation, operations, maintenance, sustainment, and modernization. These services are currently provided to U.S. government agencies, including the FAA, U.S. Air Force, U.S. Navy, U.S. Army, Defense Threat Reduction Agency (DTRA), and the intelligence community; additionally, the business includes a growing commercial, global aviation presence. Before joining Exelis, Dr. Drew was the senior vice president of Strategic Capabilities and Technology at TASC, Inc., leading an enterprise-wide team that provided systems engineering and integration, cyber security, financial and business analytics, and test and evaluation solutions to intelligence, defense, and federal and civil customers. In a prior role at TASC, she led the Enterprise Systems business unit that served defense and federal civil agencies, including DTRA, the Department of Homeland Security, and the FAA. Prior to that, Dr. Drew was sector vice president of business development for Northrop Grumman's Mission Systems sector. Before joining Northrop Grumman in 2008, she was vice president and general manager for Boeing's Integrated Defense and Security Solutions organization heading strategy and business generation in homeland and global security markets. While at Boeing, Dr. Drew also served as vice president and general manager of Boeing's C3ISR business unit serving the U.S. Air Force, U.S. Navy, and several international customers including the United Kingdom, NATO, Australia, and Turkey. In a prior role, she led a significant portion of Boeing Phantom Works developing and transitioning technology across the commercial airplane and military businesses. Dr. Drew has held several leadership roles with NRC boards and committees, including as the vice chair of the Air Force Studies Board and on the "NextGen" Air Traffic Management committee for the Transportation Research Board. She also serves on the board of directors for University of Washington's Applied Physics Laboratory. Dr. Drew has been named an associate fellow of AIAA. She also serves on the Strategic Advisory Councils to the Chancellor and Dean of Engineering at the University of Colorado, Boulder, where she earned her Ph.D. in computer science.

RICHARD V. REYNOLDS, General, U.S. Air Force (retired), is owner and principal of the VanFleet Group, LLC, an aerospace consulting company. He also serves as an independent/outside director for Allison Transmission Holdings, Inc.; Apogee Enterprises, Inc.; and Barco Federal Systems, LLC. He holds advisory board seats for Sierra Nevada Corporation and Electronic Warfare Associates-Government Systems, Inc. In a volunteer capacity, he has served as board chairman and CEO of the Air Force Museum Foundation, Inc., and as a member of the U.S. Air Force Heritage Program board of directors. In

2009-2011, he was chair of the NRC Committee on Evaluation of U.S. Air Force Preacquisition Technology Development and now serves on the Air Force Studies Board. Prior to his retirement in 2005, Gen. Reynolds was vice commander, Air Force Materiel Command. During his 34-years of active duty Air Force service, he commanded the Aeronautical Systems Center at Wright-Patterson AFB, Ohio, and the Air Force Flight Test Center at Edwards AFB, California. He was also program executive officer, airlift and trainers in the Pentagon and program director for several major weapon system acquisitions, including the B-2 Spirit. Gen. Reynolds is a graduate of U.S. Air Force Test Pilot School, Class 79B, and has more than 25 years of hands-on experience in the research, development, program management, and test and evaluation of aeronautical systems. He holds FAA certificates for airline transport pilot and flight instructor (glider), and his logbook shows more than 4,000 flying hours in 72 different military and civil aircraft. Graduating in 1971 from the U.S. Air Force Academy with a B.S. in aeronautical engineering, Gen. Reynolds has an M.S. in mechanical engineering from California State University and an M.A. in national security and strategic studies from the Naval War College. He is a fellow of the Society of Experimental Test Pilots.

HARRY M. ROBINSON is the SimLEARN National Program Manager for the Veterans Health Administration (VHA) Simulation Learning Education and Research Network (SimLEARN), which uses simulation-based clinical training for health-care providers and clinicians to increase and sustain workforce skills and improve veteran patient outcomes. As the hub for the VHA National Simulation Network, the SimLEARN National Simulation Center uses innovative and immersive training technologies in a safe learning environment to enhance diagnostic, procedural, and team communication skills to support quality care and the best possible quality of care. A veteran of the U.S. Navy, Mr. Robinson completed his active duty as the commanding officer of the Naval Air Warfare Center Training Systems Division leading over 1,100 personnel accomplishing full life-cycle acquisition of training solutions for the Navy. As a naval flight officer, he primarily flew the E-2C Hawkeye and commanded both an operational squadron and type wing. His combat experience includes strike, close air support, and air superiority missions over Iraq, Afghanistan, and the former Republic of Yugoslavia. Mr. Robinson retired at the rank of captain after 28 years of military service. Subsequently he was a senior associate with Booz Allen Hamilton, where he served as the Advanced Analytics Modeling and Simulation lead supporting Team Orlando, a collaborative alliance of governmental and nonprofit agencies, including DoD and the Veterans Administration, working to leverage simulation technology to improve employee performance. His focus was on providing live, virtual, and constructive simulation to support training solutions to improve human performance and accomplish individual and team training requirements. Mr. Robinson earned his commission through the Navy Reserve Officer Training Corps upon graduation from Pennsylvania State University in 1982 with a B.S. in computer science. He then earned an M.S. in aviation systems from the University of Tennessee and completed the Naval War College Command and Staff Course. He is currently pursuing a Ph.D. in modeling and simulation from Old Dominion University and completed a Medical Modeling and Simulation Certificate Program at the Naval Postgraduate School MOVES Institute.

MICHAEL J. ZYDA is a professor of engineering practice in the Department of Computer Science at the University of Southern California. He also directs the university's GamePipe Laboratory, which engages students in research and development of interactive games. He initiated two cross-disciplinary degree programs–a B.S. in computer science (games) and an M.S. in computer science (game development)—and doubled the incoming undergraduate enrollment of the Computer Science Department. Dr. Zyda is a pioneer in the fields of computer graphics, networked virtual environments, modeling and simulation, and serious games. His research interests include collaboration in entertainment and defense, and he has developed, for example, a game used by the Army for recruiting. He has served on numerous NRC committees advising DoD. Dr. Zyda is a national associate of the National Academies and a member of the Academy of Interactive Arts and Sciences. He received a Ph.D. in computer science from Washington University.

B

Workshop Speakers

NOVEMBER 17-19, 2014
WRIGHT-PATTERSON AIR FORCE BASE, OHIO

November 17, 2014

Air Mobility Command
Lt Gen Brooks Bash, Vice Commander, Air Mobility Command
Mr. Michael "Norm" Maloy, Chief, AMC Aircrew Training Plans and Programs, Air Mobility Command

Air Combat Command
Maj Gen James Post III, Vice Commander, Air Combat Command
Mr. Fred Van Wicklin (B3H), ACC/A3TO DMO, Air Combat Command

Air Force Life Cycle Management Center
Col Daniel Marticello, Chief, Simulators Division, AFLCMC/WNS

Air Force Requirements and Agency for Modeling and Simulation
Brig Gen Eric Overturf, Mobilization Assistant to the Director of Operations, Deputy Chief of Staff for Operations, Plans and Requirements, Headquarters U.S. Air Force

November 18, 2014

Air Force Special Operations Command
Col Steve Breeze, Chief of Training, A3/A3T

Air Force Research Laboratory
Dr. Winston "Wink" Bennett, Division Technical Advisor for Training and Assessment Research, 711 Human Performance Wing

U.S. Navy
Mr. Maynard Zettler, Director, Research and Engineering Naval Air Warfare Center Training Systems Division

National Aeronautics and Space Administration
Mr. Bimal Aponso
Mr. Jim Murphy

Lockheed Martin Mission Systems and Training
Mr. Rick Boggs, Senior Fellow

Federal Aviation Administration
Dr. Jeffery Schroeder, Chief Scientific and Technical Advisor for Flight Simulation Systems

CAE, Inc.
Dr. David Graham, Director of Technology Application

November 19, 2014

Boeing Company
Mr. Steve Monson, Chief Technologist, Technical Fellow-Simulation and Training

FlightSafety International
Dr. Nidal Sammur, Director of Engineering and Simulation

Medical Simulation
Dr. Pamela Boyers, Senior Advisor to the Chancellor and Executive Director for Clinical Simulation
Assistant Professor: Department of Surgery, University of Toledo

State University of New York, Binghamton
Dr. Frank Cardullo, Professor

Feedback from Workshop Co-Champions on Next Steps
Lt Gen Brooks Bash, Vice Commander, Air Mobility Command
Maj Gen James Post III, Vice Commander, Air Combat Command